THE PHYSICS OF POPCORN

For information contact:
Kane Miller, A Division of EDC Publishing
PO Box 470663
Tulsa, OK 74147-0663
www.kanemiller.com
www.edcpub.com
www.usbornebooksandmore.com

Library of Congress Control Number: 2020935767
Printed in China
ISBN: 978-1-68464-060-7

1 2 3 4 5 6 7 8 9 10

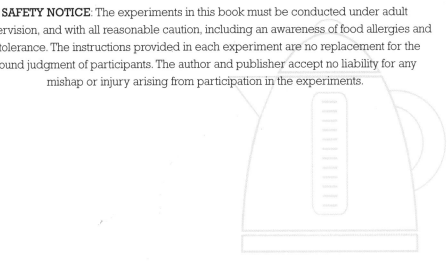

SAFETY NOTICE: The experiments in this book must be conducted under adult
supervision, and with all reasonable caution, including an awareness of food allergies and
intolerance. The instructions provided in each experiment are no replacement for the
sound judgment of participants. The author and publisher accept no liability for any
mishap or injury arising from participation in the experiments.

THE PHYSICS OF POPCORN

THE CURIOUS WORLD OF
KITCHEN SCIENCE

DR. AIDAN RANDLE-CONDE

Kane Miller
A DIVISION OF EDC PUBLISHING

CONTENTS

INTRODUCTION

On August 2, 1971, American astronaut David Scott dropped a hammer from his right hand and a feather from his left hand. They fell and hit the ground at the same time. David was standing on the moon at the time. He dropped the hammer and feather to prove an idea that was more than 300 years old. This idea came from Galileo Galilei, a sixteenth-century Italian scientist. He said that gravity—the force that makes objects fall—makes all objects fall the same way.

You would normally expect the feather to hit the ground later than the hammer because air would slow the feather down. There is no air on the moon, though, so David was able to test Galileo's idea, and prove that Galileo was right.

Even though it is important to test Galileo's idea in a place with no air, what is even more important is that the experiment worked on the moon. Galileo thought about how things fall on Earth, and came up with an idea that applies to the whole universe. Physics is about ideas that apply everywhere and at all times. Physics is about finding the most fundamental rules about how the universe works.

Physics will help you understand the world around you, and even the universe—from the tiniest bits of matter to the vast galaxies, and from the heat at the center of the sun to the chilling depths in outer space. Physics works at all times as well, so the rules that shape the universe today are the same as they were 1 billion years ago, and they will be the same in a trillion years' time.

In this book, you will be guided through the main areas of physics. You'll find plenty of experiments, including how to make space shuttles, together with explanatory text and fun quizzes to make sure you develop a good understanding of the concepts. The experiments involve everyday items; if you don't have a piece of the equipment listed, try asking a friend or a neighbor.

But what does popcorn have to do with physics? It turns out quite a lot because the physics of popcorn is the same as the physics of everything around us. For example, the physics of steam is not only involved in making popcorn; it also powered the Industrial Revolution. The waves that make a microwave oven cook the popcorn are also used in every single cell phone and radio in the world. The same kinds of atoms that popcorn is made of were created in the center of stars, and they can be found all over the universe.

You may be making everyday items today, but years from now you could be flying through space. You could be experimenting with magnets today, and later spend your career inventing incredible new machines. Once you understand physics, the only limit to what you can do are the laws of nature. By understanding these laws, humans have been able to fly to the moon; create the Internet; and generate clean, renewable energy.

CHAPTER 1
ENERGY AND HEAT

DISCOVER...

LEARN...

EXPERIMENT...

DISCOVER: DENSITY AND FLOATING

Unpopped kernels of popcorn are small and hard. When they pop, a larger, softer, tastier food is produced. What makes a kernel pop? Inside each kernel there is water and hard starch. Heat is needed to soften the starch and break open the kernel.

MASS AND VOLUME

The first thing to look at is how the density of the popcorn changes. Density states how much mass there is in a unit of volume. Mass describes how heavy an object is, and volume is a way of measuring how much space an object takes up. For example, if you have a pint of milk, the pint is the unit of volume. A heavy object has a lot of mass, and a light object has little mass. If you assume that every popcorn kernel has the same mass, doubling the number of kernels should double the mass. Another way of saying this is that the mass is proportional to the number of popcorn kernels.

The density is given by the equation:

Density = Mass ÷ Volume

That said, density does not always refer to mass. For example, energy density refers to how much energy there is in a unit of volume. Popcorn kernel density can be defined as the number of popcorn kernels in a unit of volume.

VISUALIZING EQUATIONS

Throughout this book you will be working with equations similar to that for density. They can be represented by a triangle. If you know two values, and you need to know the third, you can cover up the unknown value with your hand to find out how to calculate it.

For example, to calculate density, cover it up on the triangle below. You'll find that you need to divide the mass by the volume. If you wanted to calculate the mass, you would cover it up and find you need to multiply the density by the volume.

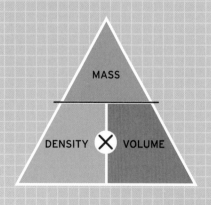

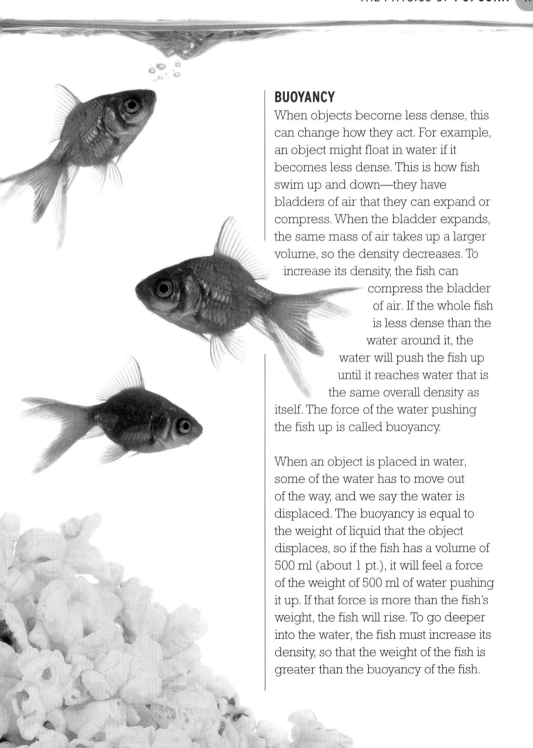

BUOYANCY

When objects become less dense, this can change how they act. For example, an object might float in water if it becomes less dense. This is how fish swim up and down—they have bladders of air that they can expand or compress. When the bladder expands, the same mass of air takes up a larger volume, so the density decreases. To increase its density, the fish can compress the bladder of air. If the whole fish is less dense than the water around it, the water will push the fish up until it reaches water that is the same overall density as itself. The force of the water pushing the fish up is called buoyancy.

When an object is placed in water, some of the water has to move out of the way, and we say the water is displaced. The buoyancy is equal to the weight of liquid that the object displaces, so if the fish has a volume of 500 ml (about 1 pt.), it will feel a force of the weight of 500 ml of water pushing it up. If that force is more than the fish's weight, the fish will rise. To go deeper into the water, the fish must increase its density, so that the weight of the fish is greater than the buoyancy of the fish.

EXPERIMENT: THE FIRST POP

When popcorn is made, one thing you'll quickly notice is how much more space popped popcorn occupies, compared to the kernels. How can you find out the volume of popcorn before and after it has been cooked?

YOU WILL NEED:

- ⅓ cup popcorn kernels
- Kitchen stove and a saucepan with lid
- Fine filler material (such as sugar, salt, rice, or sand)
- Kitchen scale and heatproof mat
- Small measuring pitcher (able to measure in 10 ml increments)
- Medium-size measuring pitcher (able to measure 750 ml)
- Large mixing bowl

ADULT SUPERVISION REQUIRED

WHAT TO DO:

1. Heat the pan on the stove at medium heat. Add the kernels, and then cover with the lid. Shake the pan every 20 seconds. Once the kernels begin to pop, shake the pan every 10 seconds. Remove the pan from the stove when half of the kernels have popped (ask an adult for help). Place the pan on a heatproof mat to cool (careful, it will be hot). Once the popcorn has cooled, separate out the unpopped and popped kernels.

2. Pour 40 ml of water into the small measuring pitcher and, using the kitchen scale, record its mass.

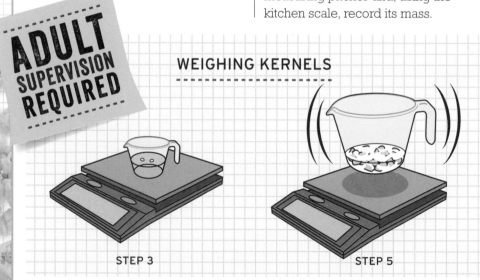

WEIGHING KERNELS

STEP 3   STEP 5

3. Add the unpopped kernels to the small pitcher one at a time until the water level rises to 50 ml. Count how many kernels you added. This tells you how many unpopped kernels take up 10 ml of space. Record this as number N_unpopped. Weigh the pitcher and its contents again.

4. Subtract the mass of the small pitcher with only water (see step 2) from the mass of the small pitcher with water and unpopped kernels (see step 3). Record this as M_unpopped. Dividing this by the number of unpopped kernels added in step 3 will tell you how much a single popcorn kernel weighs on average.

5. Place 25 popped kernels into the medium-size measuring pitcher. Add enough filler material to cover up the kernels. Shake the pitcher gently for 30 seconds, then use the markings on the pitcher to record how much volume is taken up. Weigh the pitcher, including the filler and the popped kernels, to find and record the mass.

6. Empty the medium-size measuring pitcher into the large mixing bowl, take out the unpopped kernels, and return all the filler to the pitcher. Weigh the pitcher and record the mass, and record the volume of the filler.

7. Subtract the volume of the filler (see step 6) from the volume of the popped kernels and filler (see step 5).

You now have the volume taken up by the 25 popped kernels. Record this volume as V_popped.

8. Subtract the mass of the filler (see step 6) frvom the mass of the popped kernels and filler (see step 5). You now have the mass of 25 popped kernels. Record this mass as M_popped.

WHAT HAPPENS?

How much more space does the popped popcorn take up compared to the unpopped kernels? In order to answer this, you first need to know how much space a single piece of popcorn takes up.

To calculate the average volume of an unpopped kernel, look again at step 3. Divide the 10 ml by the number of unpopped kernels you added (N_unpopped). This gives you a volume that you can call V_unpopped. Divide the volume of the popped popcorn (V_popped, see step 7) by 25 to get the average volume of a piece of popped popcorn. Find the difference by subtracting the smaller volume from the larger volume. Is the difference in volume larger than you expected? (You will use the measurements from this experiment on page 20.)

To find out where the extra volume came from, add some of the popped popcorn to water and see how much the water level increases. Does it increase by as much as you expect? If not, why do you think this is? (For an explanation, see page 148.)

DISCOVER: STATES OF MATTER

Everything you can see and touch around you is made of matter. There are three main states of matter: solid, liquid, and gas. For example, the food you eat is a solid; the water you drink is a liquid; and the air you breathe is a gas.

SOLID, LIQUID, GAS

Everything is made up of molecules, which are tiny pieces of matter. Whether something is a solid, a liquid, or a gas depends on how these molecules hold themselves to each other.

Solids are hard and keep their shape. The molecules in a solid are arranged like oranges in a box, so that if you try to move one of them, they all move. Solids are usually more dense than gases.

Molecules in liquids are like oranges in a bag. They will tend to stay together, but their shape is not fixed, and they will fill whatever container you put them in. Liquids are normally about as dense as solids.

Molecules in gases are like a bunch of oranges flying around. Gases do not have a fixed shape or density, and they will fill up any available space.

Liquids and gases are fluids. When the molecules of a fluid hit the sides of a container, they will exert pressure.

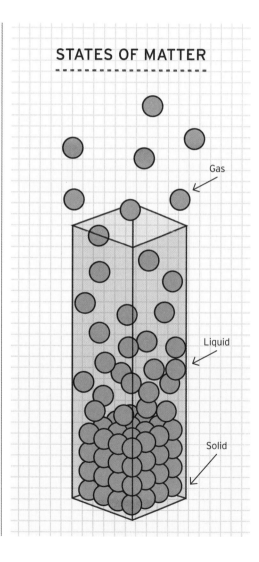

STATES OF MATTER

Gas

Liquid

Solid

GAS IN A BALLOON

Imagine you are given a balloon that has air in it, but whose opening hasn't been tied shut yet. If you gently squeezed the balloon while holding the opening closed, you would be able to feel the air inside pushing back at you. This is because of pressure. The gas molecules inside the balloon are flying around at high speed, bumping into each other and the interior of the balloon. When pressure is applied on the surface, you can feel it as a force.

What could you do to increase the pressure inside the balloon? One option would be to add more air to it by blowing it up some more. This can be written down as an equation:

Pressure inside balloon = Constant × Amount of air inside balloon

This means that if you double the amount of air inside the balloon, the pressure will double. The pressure, amount of gas, volume, and temperature are all variables, which means they can change. The constant is a number that does not change, and its value depends on the units that are used. For example, imagine you were driving from New York to Los Angeles. The distance would be 2,800 miles, but you want to know what this distance is in kilometers, so you need to multiply 2,800 by a constant of 1.6 (the number of kilometers in a mile). This gives a distance of 4,500 km.

Another thing you could do to increase the pressure is to squash the balloon. If you force it to take up a smaller amount of space, you will feel the pressure inside the balloon pushing back at you. That means that if you decrease the volume, the pressure will increase. The equation can be updated to reflect this:

Pressure inside balloon = Constant × Amount of air inside balloon ÷ Volume

One other option would be to add some heat to the balloon. When you heat something up, it usually expands and tries to take up more space. If the air inside the balloon is not able to expand, the pressure will increase instead. This means that one more part can be added to the equation:

Pressure inside balloon = Constant × Amount of air inside balloon × Temperature ÷ Volume

DISCOVER: HEAT AND ENERGY

Energy is all around you. It's in the rays of the sun, and it's used by the cars being driven on the streets. Everything humans do, even breathing, relies on energy, and humans seem to use more and more of it over time.

Energy is used to make popcorn pop. It is a conserved quantity, which means that it cannot be created or destroyed—it can only be transferred from one kind of energy to another. Energy can exist in many forms, and one of them is heat. Heat can be converted into work (which is what happens in a power plant), and work can be converted into heat—which is what happens in an electric heater, or if you rub your hands together quickly.

Every time energy is converted from one form into another, some of the energy is wasted in the form of heat. This is because everything gets less well ordered over time. Think about how hard it is to keep a room tidy. You can see this kind of waste heat being generated in a lot of places. Computers use electrical energy to perform calculations, and they need to have cooling fans to remove the waste heat. When you charge a cell phone, you can usually feel it getting slightly warmer. After riding a bicycle, the tires, brakes, and even the gears are warmer.

MOVING HEAT AROUND

If there is a lot of heat in a small space, it will tend to spread out and fill a bigger space over time. If you get into a hot bath, you will notice the temperature decreases quickly when it is very hot. Later on, when the water is only slightly warmer than room temperature, it will cool down more slowly.

You can feel the heat coming off a hot drink through conduction (the cup is hot) and convection (the air above it is hot).

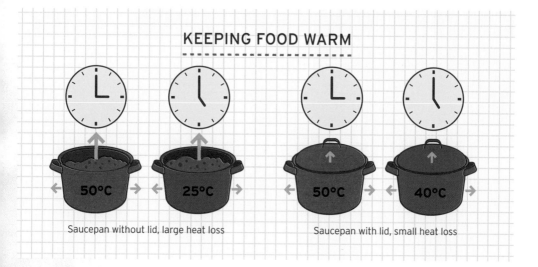

KEEPING FOOD WARM

← 50°C → ← 25°C →

Saucepan without lid, large heat loss

← 50°C → ← 40°C →

Saucepan with lid, small heat loss

Heat is transferred in three ways:
• Conduction: heat transferred through a solid
• Convection: heat transferred through a liquid or gas
• Radiation: heat transferred from one place to another, without needing a solid, a liquid, or a gas

When popcorn is heated, all three of these forms of heat transfer take place. Since heat is a form of energy, and energy is conserved, heat is conserved. A hot object will lose heat, and if heat moves out of a hot object quickly, it will cool down quickly. If heat moves out slowly, the object will cool slowly. That is why hot water tanks are covered with insulation—because insulation slows down heat loss.

The speed, or rate, of heat loss relates to how fast heat moves. The bigger the difference in temperature, the faster the heat will be transferred.

COOLING BY HEATING

When ice is added to a drink, it might be thought that the ice cools the drink around it. But is that what really happens? The water is warmer than the ice, and its heat is transferred to the ice. As the ice melts, it produces more water, which helps the convection process deliver yet more heat to the ice. As the water heats the ice, it loses heat energy and gets cooler. (It turns out that adding heat to the ice doesn't just make the ice warmer; it causes it to melt. Most of the heat is lost through the *melting* of the ice, rather than the warming of the ice.)

EXPERIMENT: MODEL OF CONVECTION

If you turn a radiator on in a cold room and wait for the room to warm up, you will soon notice how heat moves around. The closer you are to the heater, the warmer the air will feel, compared to elsewhere in the room. Try this experiment to help you visualize how heat moves.

YOU WILL NEED:
- 2 glasses that are the same size and shape
- Tap water
- Water-based red and blue food coloring
- Stopwatch

WHAT TO DO:
1. Fill the glasses three-quarters of the way full with water.

2. Add a single drop of blue food coloring to one of the glasses (this will be glass 1) and time how long it takes before the food coloring changes the color of the water. Stop the timer when the water is a single, even shade of blue. While the timer is running, observe how the food coloring starts

Water-based food coloring is best.

STEP 2

off concentrated in one small region of the water and spreads out to fill all available space. You'll also notice how the coloring spreads quickly at first and then increasingly slowly. Leave the blue water in the glass.

3. Now add a drop of blue food coloring to glass 1 and glass 2 at the same time. In which glass does the water take the longest to become a single color? How many seconds did it take?

4. Leaving the water in the glasses, add a drop of red food coloring to glass 2 and time how long it takes to spread out. How long did it take to reach a single shade of purple?

DISCOVER MORE

By the end of step 2 in this experiment, the food coloring in the glass of water has become evenly distributed. You might think that this means the coloring has stopped moving, but this isn't true. It moves around in the water constantly, but because the color density is the same throughout the liquid, there is no change in the color of the water. Even though the water might look peaceful, it is full of motion.

WHAT HAPPENS?

This process of spreading out is known as diffusion. In steps 2 and 4, the diffusion of the food coloring occurs at the same speed, or rate. Even though red is added to blue in step 4, the rate of diffusion doesn't change. However, a second drop of the same color will take longer (as you saw with glass 1 in the third step).

Diffusion takes place more quickly when the differences in color density are larger. This makes sense when you think about it. When you've eaten hot food—for example, a hot soup—you may have noticed how quickly it cools down at first. If you leave the soup out for a long time, it will cool down ever more slowly, so that the change in temperature becomes less and less obvious.

STEP 4

LEARN ABOUT: DENSITY

Scientists want to understand the world around them. Sometimes the things they want to know are not so easy to measure, and they have to find the answer indirectly.

POP QUIZ: POPCORN DENSITY

In this activity, you will calculate the density of popcorn using the measurements you calculated in the experiment on page 12.

To calculate the density of popcorn in its unpopped and popped forms, consider each state. For the unpopped state, you know how much volume the kernels occupied (as you discovered on pages 12-13, this is the change in the level of the water, from 40 ml to 50 ml). You also know the mass of the kernels. Using the equation for density in terms of mass and volume (see page 10 for a reminder), calculate the density of the unpopped popcorn:

D_unpopped = M_unpopped ÷ V_unpopped

For the popped popcorn, you use the same method. You know how much mass the popped kernels had, and how much volume they occupied, so you can calculate the density of the popped kernels:

D_popped = M_popped ÷ V_popped

To find out how much more space popped popcorn takes up compared to the unpopped kernels, divide D_popped by D_unpopped. How much smaller is the density of the popped popcorn compared to the unpopped kernels?

Perhaps the popcorn lost some mass when it popped. Take the mass of the unpopped kernels and divide it by the number of unpopped kernels to get the average mass per kernel. You can do the same for the average mass per kernel of the popped popcorn. The values should be roughly the same, and it's certainly not enough to explain why the unpopped kernels are so much more dense than the popped popcorn.

How much did the mass per kernel of popcorn change after popping? If you were in charge of making and selling popcorn, would you rather transfer it in its unpopped or popped state?

LEARN ABOUT: HEAT TRANSFER

Heat can be transferred in a solid by conduction, in a liquid or gas by convection, and across empty space by radiation.

POP QUIZ: HEATING UP

Which processes transfer heat in the following examples? (More than one process may apply in some cases.)

1. It's time for dinner, and your friend wants to make pasta. They put some pasta in a pot of cold water and cook it on the stove.

2. It's a hot day, and you're drinking lemonade. A piece of ice falls out of your glass and onto the ground. The ice is melted by the sun.

3. You're eating chocolate, but you are taking your time. It melts in your hands.

4. You start to run a hot bath, and you can see water vapor rising from the surface of the water.

5. You're camping, and you warm yourself by a campfire.

6. You sprain your ankle, and the school nurse brings you an ice pack, which you apply to the sprain.

7. Your parents drive you to school on a cold day. They turn on the heater in the car to blow warm air into the car.

Chocolate will melt if you hold it in your hand too long.

DISCOVER: IDEAL GASES AND STEAM ENGINES

One of the discoveries that revolutionized human understanding of the world is that heat is a form of energy. Heat is actually the energy from molecules moving around, which is called kinetic energy. The more the molecules move around in a substance, the more heat the substance has.

The temperature of a substance relates to how much the molecules are moving around. Heat moves from regions of high temperature to regions of low temperature. Adding heat to a substance increases its temperature—by how much depends on the substance. If you add the same amount of heat to water and vegetable oil, the oil's temperature will increase more quickly because water can hold more heat. Of all the liquids encountered on a daily basis, a cup of water can hold the most heat for a given change in temperature. It takes a lot more heat energy to warm up water than oil.

When heat is added to a gas, the molecules inside it move faster. As the molecules bounce around, the gas will usually expand. If it can't expand because it's contained within a vessel, the molecules will hit the side of the vessel with a higher speed, and the pressure will increase.

THE EFFECT OF HEAT ON WATER MOLECULES

COLD WATER

WARM WATER

HOT WATER

THE IDEAL GAS EQUATION

The way the temperature, volume, and pressure of a gas are related is described by the ideal gas equation:

Pressure × Volume = Constant × Amount of gas × Temperature

This is similar to the equations on page 15. As you can see, it contains a constant value, which is just a number. If one person measures temperature in Fahrenheit, and another person measures temperature in Celsius, they would need to use different values for the constant. The constant doesn't affect the physics, though. What is important is that the pressure and volume on one side of the equation balance the temperature and amount of gas on the other.

If the temperature of a gas is increased, so must the pressure or volume, or pressure *and* volume, increase.

What if the gas can't increase its volume? In that case, the pressure of the gas must increase. This is how a steam engine works—the pressure of the steam is used to push a piston.

USING HEAT TO DO WORK

The process of using heat to make steam to generate power changed the world during the Industrial Revolution. It made steam trains possible, as well as other machines that have made people's lives easier. How does a steam engine convert heat into work? First, water must be heated enough to produce high-pressure steam inside a closed vessel. When pressure is exerted on a surface, it results in a force. The pressure, force, and area are related by the equation:

Force = Pressure × Area

By adding heat to steam in a closed vessel, the pressure increases, and if this pushes on a surface that can move (in this case, a piston), then the steam can be used to do work. But how much work? When a force is moved across a distance, it does work. This is what happens when you pick up a heavy object. You have to lift the object against gravity, and that takes effort. In a similar way, when there is a tug-of-war competition, both teams pull against each other. Every time the rope moves one way or the other, some work is done. Work is given by:

Work = Force × Distance

This way, a steam engine can be used to convert heat energy into useful work, by way of pushing a piston.

001 252

EXPERIMENT: TURNING WATER TO STEAM

When heat is added to water, the water molecules vibrate and bump into each other at higher speed. If a molecule receives enough energy, it can escape as steam, which is what happens when water boils. What happens when water is boiled in an electric teakettle? In particular, how much space does steam take up, compared to water?

YOU WILL NEED:

- Measuring pitcher (able to measure in liters)
- Stopwatch
- Digital kitchen scale
- Small glass
- Electric kettle
- Ruler (able to measure in centimeters)
- Sticky notes

ADULT SUPERVISION REQUIRED

WHAT TO DO:

1. Position the kettle in front of a wall. Place a sticky note on the wall 50 cm above the level of the kettle's spout.

2. Record in centimeters the width of the spout.

3. Add 1 L of water to the kettle.

4. Making sure that the kettle is unplugged, record the mass of the kettle using the kitchen scale.

5. Move the kettle off the scale, plug it in, and turn it on.

6. As it heats, you will see steam appear. When the first stream of steam begins to escape, start the stopwatch.

7. As the steam escapes, estimate how long it takes to rise 50 cm to the marker. (Look at the swirling shapes in the steam and follow one from the spout to the marker. You may want to repeat the experiment in order to perform this step again.)

8. When the water boils and the kettle switches itself off, stop the stopwatch.

9. Unplug the kettle and record its mass a second time (be careful; it will be hot). Subtract this mass from the mass in step 4 to find the mass of the water lost.

10. Place the glass on the scale and set the scale back to a weight of zero.

11. Add enough water to the glass so that the mass reading on the scale is equal to the mass of the water lost. (You calculated this in step 9.)

12. Pour this into the measuring pitcher to see what volume of water was boiled to make steam.

ANALYSIS

You now need to work out the volume of steam the kettle produced. The formula you'll use is:

Volume of steam = Area of spout × Speed of steam × Time taken to boil

You know the mass of the water lost. This is equal to the mass of the steam produced. You now need to calculate the area of the spout, which you can assume is a perfect circle. The equation for the area of a circle is π × Diameter × Diameter ÷ 4. Pi (π) is a number that is roughly 3.14, and the diameter is the width of the circle (see the diagram). If the diameter is 3 cm, the area would be:
$3.14 \times 3 \text{ cm} \times 3 \text{ cm} \div 4 = 7.1 \text{ cm}^2$.

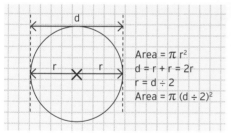

$Area = \pi r^2$
$d = r + r = 2r$
$r = d \div 2$
$Area = \pi (d \div 2)^2$

Next, estimate the speed of the steam escaping the spout. You can use the following equation to help you:

Speed = Distance ÷ Time

Read the time taken for the water to boil from the stopwatch. Now that you've figured out all three values, you can determine the total value of the steam produced by multiplying the speed of the steam by the area of the spout (this is the volume of steam produced per second), then multiplying that value by the time taken for the water to boil.

How does it compare to the volume of the water lost? You should find that the volume of the steam was hundreds of times larger than the volume of water lost.

WHAT HAPPENS?

When water boils, it makes steam. Steam takes up a very large amount of space compared to water. The answer is only approximate because there were several assumptions. Is the spout really a perfect circle? Does the steam always move at the same speed? Is the amount of steam produced constant? How easy was it to estimate how fast the steam rose? Every one of these factors will change the answer slightly.

DISCOVER: POPPING THE KERNEL

Pages 24–25 showed that steam takes up hundreds of times as much volume as water when water is heated to 100°C (212°F). The volume of water in its liquid state doesn't change much as it is heated, increasing by only about 4% when heated from 0°C (32°F) to 100°C. What happens if you make steam and it cannot expand?

Steam is a gas, and it mixes with the air around it. This can be described by the ideal gas equation (see page 23):

Pressure × Volume = Constant × Amount of gas × Temperature

Think about what this equation tells you about steam inside a popcorn kernel at 100°C (212°F). The kernel is a closed vessel, so the volume stays the same. The temperature is 100°C, so that stays the same. That means if you increase the amount of steam by boiling water, the pressure of the steam must increase.

In the Turning Water to Steam experiment, when the water boiled, it changed state, from liquid to gas, and the pressure stayed the same. (The pressure was the same as the pressure in the rest of the room because the kettle is an open vessel—steam can escape from the spout.)

THE KERNEL POPS

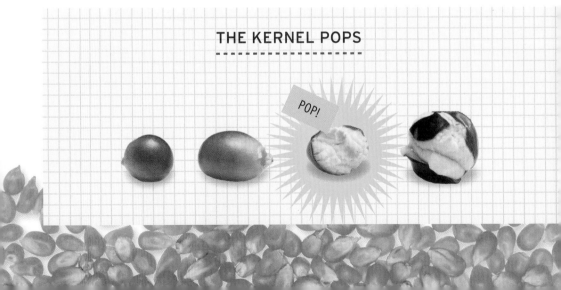

POP!

What would happen if the volume of the air and steam mixture were kept the same? Or put another way, what would have happened in your experiment if the kettle were sealed, with no spout to release steam? Instead of the volume of steam increasing, the pressure of the steam would increase. If the kettle wasn't strong enough, it would rupture from the pressure. So, it's good that kettles have spouts.

PRESSURE COOKER

When popcorn is heated, the same process takes place inside the kernel. When the small amount of water in the kernel is heated, it boils to produce steam. Increasing the amount of steam increases the energy inside the kernel. The pressure of the steam builds up until the kernel can no longer withstand the pressure. At the weakest point of the kernel, the surface is torn open, and the popcorn expands with a pop. At this point, the steam is able to expand very quickly, and this causes the starch and protein in the popcorn to expand rapidly, changing from a hard, dense material to a light and soft snack.

EXPERIMENT: THE PRESSURE OF THE POP

The hull of a popcorn kernel is very hard and difficult to break open. Can you find out how much pressure it takes to make the kernel pop? Use the results from your previous popcorn experiment to help you find the answer.

YOU WILL NEED:

- 1 bag of microwave popcorn
- Microwave oven
- Kitchen scale (able to measure in grams)

WHAT TO DO:

1. Take a bag of uncooked microwave popcorn and weigh it using the kitchen scale.

2. Cook the popcorn according to the instructions. When it's cooked, carefully open the bag to allow the steam to escape. (It will be hot, so ask an adult to help.)

3. Record the mass of the full bag.

4. Empty the bag and separate the popped and unpopped kernels. Record the mass of the empty bag.

RELEASING THE STEAM AND COUNTING THE KERNELS

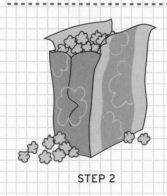

STEP 2

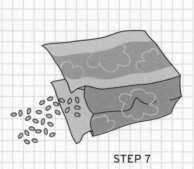

STEP 7

Since you know the mass of the empty bag, you can find out the mass of the popcorn before and after cooking.

5. Subtract the mass of the empty bag from the mass of the uncooked popcorn to find out the mass of the popcorn before cooking. Similarly, subtract the mass of the empty bag from the mass of the cooked popcorn to find out the mass of the popcorn after cooking. The difference of these two values is the mass of the steam that escaped.

7. Count out how many kernels remained unpopped and then—assuming each unpopped kernel weighs 0.16 g—calculate their total mass. Subtract this value from the mass of all the popcorn after cooking, and you will have the mass of the popped popcorn.

8. Assuming the average mass of a popped kernel is 0.14 g, calculate the number of pieces of popped popcorn you have.

UNDER PRESSURE

You now know how much steam was generated, and how many kernels were popped, so you can find how much steam was produced inside a kernel. Now, you need to know the volume of an unpopped kernel. Since measuring such a small object is quite hard, you can line up 20 kernels in a row, measure the total width of them in centimeters, and divide by 20. To find the volume, you can use the equation for the volume of a sphere:

> **Volume of a sphere =**
> $4 \times \pi \times \text{radius}^3 \div 3$

For example, if you measured 0.5 cm as the radius of the popcorn kernel, then its volume would be:

> $4 \times \pi \times (0.5 \times 0.5 \times 0.5) \div 3$
> $= 4 \times 3.14 \times 0.125 \div 3 = 0.52 \text{ cm}^3$

As you have seen, the ideal gas equation is Pressure × Volume = Constant × Amount of gas × Temperature. Popcorn usually pops at around 176°C (350°F), so you know the temperature. This equation gives an answer compared to the pressure of air at sea level.

> **Pressure = Amount of steam per kernel (g) × Temperature (C) ÷ Volume of kernel (cm³)**

Put the known values into the equation to see how much pressure it takes to break open a popcorn kernel. (See page 149 for an example calculation.) If your value for the pressure comes out to 2, that means that the pressure inside the kernel is two times larger than the pressure of air around you (if you live at sea level). Air pressure is lower at higher altitudes, so if you live at a high altitude—for example, in Denver—you can expect the air pressure around you to be about 20% lower than at sea level.

LEARN ABOUT: BALANCING EQUATIONS

Physics is all about making equations balance, and keeping equations balanced tells you how different values relate to each other. Understanding this will allow you to answer questions such as, "If the speed of a car increases, what else will change?"

CHALLENGING EQUATIONS

In 2009, sprinter Usain Bolt broke the world record for the 100 meters. He ran 100 m in 9.58 seconds. His average speed in the race can be calculated by the equation:

Speed = Distance ÷ Time

So, 100 m ÷ 9.58 sec = 10.4 m/s, meaning that Bolt ran at an average speed of 10.4 m/s.

The previous world record was set by Tyson Gay, who ran 100 m in 9.69 seconds. What was Gay's average speed?

SPEED, TIME, AND DISTANCE
To see how speed, distance, and time are related, you can make a seesaw diagram:

Distance = Speed x Time

When the seesaw is balanced, the left-hand side is equal to the right-hand side. In the example of Tyson Gay, he ran 100 m, so you can put 100 m on the left-hand side. On the right-hand side, the time was 9.69 seconds, and the speed was 10.3 m/s.

Distance = Speed x Time
100 meters = 10.3 m/s x 9.69 seconds

Usain Bolt ran the same distance in less time. This means that the left-hand side remains the same and the time on the right-hand side decreases. If the seesaw is to stay balanced, the speed must therefore increase.

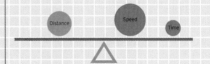

Distance = Speed x Time
100 m = 10.4 m/s x 9.58 seconds

PRESSURE, VOLUME, TEMPERATURE, AND GAS

The seesaw diagram can also be applied to gases, using the formula:

Pressure × Volume = Constant × Amount of gas × Temperature

Volume x Pressure = Amount of gas x Temperature

This lets you see what happens when a gas is heated in a metal cylinder. As it heats up, the temperature increases:

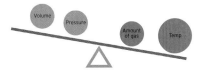

Volume x Pressure = Amount of gas x Temperature
Temperature increases, seesaw is unbalanced.

This makes the seesaw tilt to the right. To make it balance, either something on the left-hand side must increase, or something on the right-hand side must decrease. The volume must stay the same because the gas is in a cylinder, and heating the gas doesn't change how much gas there is. That means that the only thing that can change is pressure. Therefore, the pressure increases to balance the seesaw.

Volume x Pressure = Amount of gas x Temperature
Temperature increases, pressure increases, seesaw is balanced.

There are four seesaw diagrams below. See if you can match them to their descriptions.

- A popcorn kernel is heated. The amount of steam (which is a gas) inside it increases, and so does the temperature. The volume of steam stays the same.
- The air inside a hot-air balloon cools down. The pressure of the air does not change, and the amount of air stays the same.
- You blow up a balloon and add more air to it. The temperature of the air stays the same.
- You open a bottle of soda and notice a lot of air escapes from the bottle. The temperature of the air stays the same.

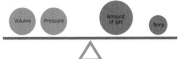

Volume x Pressure = Amount of gas x Temperature
Temperature and amount of gas increase, pressure increases, seesaw is balanced.

Volume x Pressure = Amount of gas x Temperature
Amount of gas decreases, pressure decreases, seesaw is balanced.

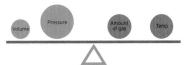

Volume x Pressure = Amount of gas x Temperature
Amount of gas increases, pressure and volume increase, seesaw is balanced.

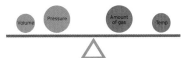

Volume x Pressure = Amount of gas x Temperature
Temperature decreases, volume decreases, seesaw is balanced.

EXPERIMENT: FLY A HOT-AIR BALLOON

Gases can act in very strange ways when they are heated up or cooled down. For centuries, people have used this knowledge to make vehicles that float in air and water. Heating air makes it expand, which can be used to make hot-air balloons fly.

YOU WILL NEED:

- Large plastic shopping bag with no holes in it
- 6 paper drinking straws
- Sticky tape
- Scissors
- Hair dryer
- 2 balloons
- Freezer

MAKE A HOT-AIR BALLOON

A hot-air balloon is made of two main components: a large sack that contains the air, which is called the envelope, and a way to heat the air in the envelope, which is called the heat source. The shopping bag will be used as the envelope, and the hair dryer will be used as the heat source. The opening of the shopping bag needs to be held open so that the hair dryer heats the air effectively.

WHAT TO DO:

1. Using the sticky tape and scissors, tape drinking straws along the opening of the shopping bag, with no space between the straws. This will in effect turn the "mouth" of the bag into a polygon and ensure that the bag does not close by itself.

2. Hold the plastic bag upside down with one hand. With the other hand, hold the hair dryer below the opening of the hot-air balloon, pointing up toward the ceiling.

3. Set the hair dryer to its warmest setting. You should see the bag start to expand. If you release the hot-air balloon, it will start to rise and float. How long can you keep the hot-air balloon afloat with the hair dryer? How easy is it to keep the hot-air balloon afloat with the hair dryer on the coolest setting?

ADULT SUPERVISION REQUIRED

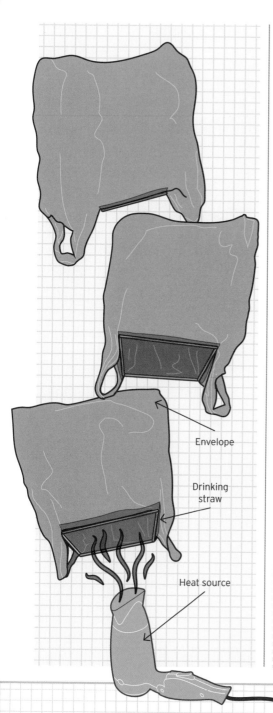

Envelope

Drinking straw

Heat source

FREEZING A BALLOON

Gases act differently at different temperatures.

WHAT TO DO:

1. Blow up two balloons until they are about the same size, and tie them closed. Label them with a marker, or use two balloons of different colors.

2. Put one balloon in the freezer for 4 hours, and keep the other balloon somewhere secure.

3. After 4 hours, take the balloon out of the freezer and compare its size to that of the other balloon. Leave both balloons at room temperature for an hour and compare their sizes again.

WHAT HAPPENS?

When gases are heated, the temperature increases. This makes the volume increase to balance the equation. When the air in the bag expands, it has a lower density, which causes it to rise. It's easier to make the bag float with the hair dryer on the warmer settings because this heats up the air in the bag faster.

When gases are cooled, the temperature drops and the volume decreases, so the balloon in the freezer will be smaller than the balloon at room temperature. As the air in the balloon from the freezer warms up, its volume will increase, and both balloons will eventually be the same size.

LEARN ABOUT: PROPERTIES OF SOLIDS, LIQUIDS, AND GASES

Water is very versatile because it can exist in solid, liquid, and gaseous states. Even though they are all forms of water, the different states act very differently. The properties of water in its three states make it very useful for activities such as cooking food.

POP QUIZ: SOLIDS, LIQUIDS, OR GASES?

Can you work out whether these properties match up best to solids (s), liquids (l), or gases (g)? The properties are arranged into groups of three, with one for solids, one for liquids, and one for gases. Each group describes how solids, liquids, and gases act differently.

1. a) The molecules have no overall pattern and don't tend to stay together.

1. b) The molecules are well organized in a repeating pattern.

1. c) The molecules have no overall pattern, but are generally close together.

2. a) The molecules touch each other.

2. b) There is a lot of space between the molecules.

2. c) The molecules are close together, but can slide over each other.

3. a) The molecules can only vibrate in place. They cannot move around.

3. b) The molecules are free to move around each other, but tend to stay close to other molecules.

3. c) The molecules are free to move anywhere, and move quickly in all directions.

4. a) Examples include water, oil, and lava.

4. b) Examples include air, steam, and helium.

4. c) Examples include sand, wood, and rubber.

5. a) The overall shape matches the shape of its container, but its volume stays the same.

6. a) When heated, it eventually boils.

5. b) The overall shape stays the same unless it gets deformed.

6. b) When heated, it expands or its pressure increases.

5. c) The overall shape matches the shape of its container, but its volume will expand to fill its container.

6. c) When heated, it expands slightly and eventually melts.

7. a) If you hit it with a hammer, nothing significant happens.

7. b) If you hit it with a hammer, only the substance that touches the hammer will move. You might make waves appear.

7. c) If you hit it with a hammer, the whole substance will move.

CHAPTER 2
WAVES AND
ELECTROMAGNETISM

DISCOVER...

LEARN...

EXPERIMENT...

DISCOVER: ELECTRICITY AND MAGNETISM

Electricity and magnetism are used all the time in daily life. Electricity powers homes, makes the lights work, and is used in all kinds of devices. Magnetism is used in any kind of motor or power generator. It's also what keeps refrigerator doors closed.

Electricity and magnetism are two parts of the same phenomenon, which is called electromagnetism. It took a long time for scientists to discover that this was one single phenomenon. Objects can have an electric charge, which can be positive or negative. The electric force is the force felt between two objects with an electric charge. If the charges are the same (both positive or both negative), they repel (push against) each other. If the charges are opposite (one positive and one negative), they attract each other. If one of the objects has no charge, they do not repel or attract each other.

ELECTRICITY IN ACTION: LIGHTNING

When a charged object moves from one place to another, its energy can change because it feels the effects of other charged objects around it.

EARTH'S MAGNETIC FIELD

For centuries, people have used compasses to find their way around. A compass lines up with the magnetism of Earth. This magnetism comes from huge amounts of magnetic liquid rock in the center of Earth. As Earth rotates, this liquid moves around, making our planet like a giant bar magnet. The north and south poles of this magnet are close to the Antarctic and Arctic Poles, respectively. This is why the poles of magnets are named after the poles of Earth.

For example, when lightning strikes the ground, a very large number of charged particles move from the clouds, where they have a lot of energy, to the ground, where they have little energy. The energy is released, giving off a lot of light energy (lightning), sound energy (thunder), and heat energy. The heat is sufficient to start a fire if the lightning strikes a tree.

How much energy is released when lightning strikes? That will depend on a lot of different factors, but the change in energy can be measured in potential difference, which is also known as voltage. The potential difference is the amount of energy released when a charged particle moves from one place to another. When you use a toaster at 110 V, every electron that travels through the toaster can release up to 110 units of energy. This unit of energy is called an electron volt, and it is very small. It takes a lot of electrons to release enough energy to toast the bread.

WHAT MAKES MAGNETS WORK?

When a charged object moves, it creates a magnetic field around itself. The object has a north pole and a south pole, which act like the positive and negative electric charges. Any other magnetic object will respond to the magnetic field, and it will try to align itself so that its north pole points toward the south pole of the other object. This is what happens when you use a compass. Earth has its own magnetic field and the magnetic south pole is near the geographic North Pole. If you take a compass and shake it gently, or spin it around, it will always go back toward north again.

EXPERIMENT: ELECTRICITY AND MAGNETISM IN THE KITCHEN

Electricity and magnetism can be used to make things move around. Think of all the gadgets and machines that you use every day— nearly every one of them uses electricity or magnetism in some way. Static electricity and magnets are the easiest way to see the effects of electricity and magnetism around us.

PART 1: STATIC ELECTRICITY
YOU WILL NEED:

- Balloon
- Stainless steel kitchen sink, or large metal tray
- Various fabrics (felt, polyester) and hair
- Tissue paper
- Scissors
- Ruler (able to measure in centimeters)

WHAT TO DO:

In this experiment, you will compare the ability of different fabrics to generate static electricity.

1. Cut the tissue paper into equally sized small squares, around 1 × 1 cm.

2. Make sure the kitchen sink is completely dry and then place each piece of tissue paper into the sink. This ensures there is no electric charge left

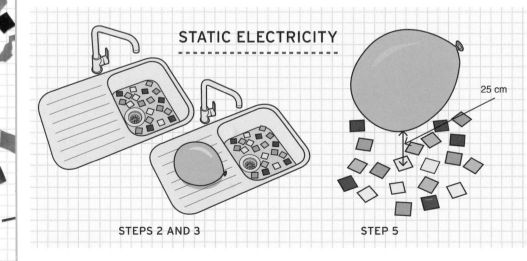

STATIC ELECTRICITY

25 cm

STEPS 2 AND 3

STEP 5

over on the pieces of tissue paper—they are now grounded with respect to each other, which means they all have the same amount of electric charge per unit area.

3. Blow up the balloon and rub the surface of it on the sink to ground it.

4. Move the pieces of tissue paper to the table and spread them out into an area roughly the same size as the balloon, but without any pieces overlapping.

5. Take the balloon and place it 25 cm above the tissue paper. Lower the balloon gradually until it is 2.5 cm above the table, and see if any of the tissue paper is attracted to the balloon.

None of the pieces of tissue paper should move because the tissue paper and the balloon were grounded with respect to each other on the kitchen sink. However, many charges per unit area are on the balloon, and the same amount are also on the pieces of tissue paper. It doesn't matter if the charges are positive or negative.

6. For the next step, rub the balloon across your hair five times. Gradually lower the balloon again from a height of 25 cm above the pieces of tissue paper to 2.5 cm. You should find that some pieces of tissue paper jump up and attach themselves to the balloon.

STEP 6

7. Write down the distance at which the first piece of tissue paper jumped up, and also count how many pieces of tissue paper are attached to the balloon.

8. Repeat this with the other fabrics you have, grounding each in the sink first. You could try rubbing the balloon against felt, polyester, carpet, or cotton. For each one, record how many pieces of tissue paper were attracted to the balloon, and at what point the first piece was attracted to it. Different fabrics will act differently, and if you used a new balloon for each fabric, you should be able to attract pieces of tissue paper from one balloon to another, if the balloons were rubbed against different fabrics.

WHAT HAPPENS?

When you rub the balloon across the fabric, some electric charges move from the fabric to the balloon. That means that the balloon has an overall negative charge compared to the tissue paper, and the tissue paper has an overall positive charge compared to the balloon. These are opposite charges, so they attract each other, and that's why the tissue paper jumps up to the balloon.

PART 2: MAGNETIC FIELDS
YOU WILL NEED:
- **3 bar magnets**
- **Piece of paper or thin cardboard**
- **Iron filings (these can be bought at a stationery store)**
- **4 books**
- **Glass jar or mug**

WHAT TO DO:
1. Arrange the books into two stacks of two, and then place the piece of paper or thin cardboard so that you make a bridge between the two stacks. The paper should be pulled taut; if necessary, place some of the books on top of the paper to hold it in place.

2. Place a bar magnet underneath the center of the paper. Tip the iron filings onto the paper. Tap one side of the paper to agitate the iron filings, and see what happens. What shape do the

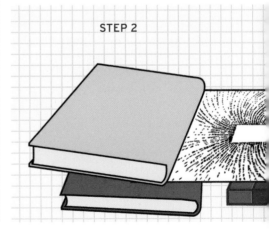

STEP 2

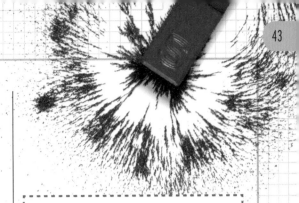

iron filings form? What happens if you place two magnets under the paper, one next to each stack of books? Experiment with the different shapes of magnetic field you can make.

3. Turn the glass jar upside down on a table and place one of the bar magnets on top of it. Hold the other two bar magnets, one in each hand, at the same height and parallel with the first magnet. Move them far enough away that the magnet on the jar will stay on the jar, but close enough that if you move them from side to side, the magnet on the jar will turn slightly. Lift the magnets in your hands about 15 cm (6 in.) and turn one of them around. Move the two magnets up and down, so that when one of them is level with the jar, the other is 15 cm above it. Can you make the magnet on the jar spin?

WHAT HAPPENS?

The magnets create their own magnetic fields, and the iron filings line themselves up with these fields, in the same way that the needle of a compass points north. By moving the magnets around, you can create different magnetic fields and get different patterns of iron filings.

This is easier to see with the three magnets. By moving the magnets around, it's possible to make the magnet on the jar act as though the North Pole keeps moving around, and it will keep moving around to point in the right direction.

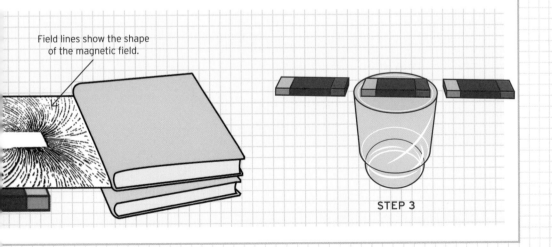

Field lines show the shape of the magnetic field.

STEP 3

LEARN ABOUT: ELECTRICITY AND MAGNETISM

You can visualize the way in which electricity and magnetism affect objects by thinking about electric and magnetic fields. The diagram below has lines that show how an electrically charged object or magnetic object will respond to electric and magnetic forces and in which direction it will move.

ELECTRIC FIELDS

As you've recently learned, charged objects are attracted to other charged objects of the opposite electric charge, and repelled by other charged objects of the same electric charge. This causes charged objects to move around if there are other charged objects nearby. Lines can be drawn from one charge to another (positive to negative) to show how objects would move. Positive charges will move with the arrows, and negative charges will move against the arrows.

In the diagram below, you can see what is known as a dipole, which is a positively charged object close to a negatively charged object. The electric field is drawn in with field lines, which go from the positively charged object to the negatively charged object. Field lines never cross each other, and the more field lines there are in a given area, the bigger the effect of the electric force.

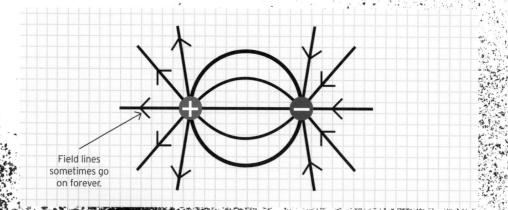

Field lines sometimes go on forever.

POP QUIZ: MAGNETIC LINES

1. These diagrams show bar magnets (the rectangles) and electrically charged objects (the circles). The field lines move from north pole to south pole (magnets), and from positive to negative (electrically charged objects). Label the north and south poles on the magnets and the electrical charge on the objects.

2. In these diagrams, the charges of the objects are there, but the field lines have been left out. See if you can fill them in, remembering to go from positive to negative. In each diagram there is a star. If you placed a positively charged object where the star is, in which direction would the object be pulled by the electric force?

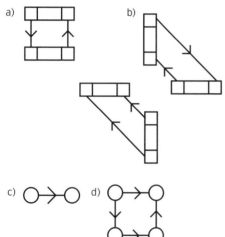

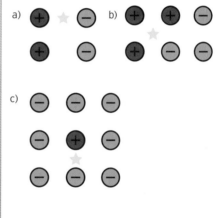

3. These diagrams have the north and south poles marked, but no field lines. Add the field lines. In each diagram there is a star. If you placed a compass where the star is, which direction would the needle face?

DISCOVER: WHAT IS A WAVE?

A wave is a regular movement of a substance or energy. That substance could be air or water; it could be the movement of a string or a solid object, or it could be in the form of sound or light. Waves are all around us, and some of them can't be seen.

A sine wave is a continuous line that goes up and down smoothly. A variable is a number that can have different values. For example, the time of day or the speed of a car can be a variable. A sine wave can be described with a few variables.

One such variable is frequency. The frequency of a wave describes how often a point moves from a peak to a valley, and back to a peak again. The speed of a wave is related to its wavelength and its frequency:

Speed = Wavelength × Frequency

The frequency describes how quickly the wave moves up and down (called oscillation). The wavelength describes the distance between two peaks or two valleys. Waves can have many different speeds and appear in a variety of situations. Waves can also be used to transmit information very efficiently— for example, sending information to and from your cell phone.

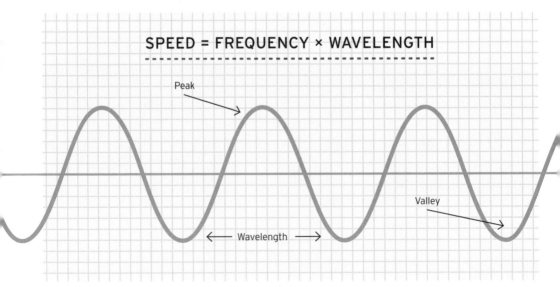

SPEED = FREQUENCY × WAVELENGTH

Peak

Valley

←— Wavelength —→

WAVES AND MUSIC

Musical instruments create sound waves, which are vibrations made in the air. When the air vibrates at a high frequency, the sound (known as pitch) is high, and when the air vibrates at a low frequency, the sound is low. An instrument like a guitar makes the air vibrate by the movement of its strings.

HARMONICS

When the string is plucked, vibrations of all different wavelengths are produced in the string. They move up and down the string, making it move to and fro by small amounts. Vibrations with wavelengths that fit perfectly along the string will move the string to and fro in the same way as they move up and down the string, reinforcing the movement. Vibrations with other wavelengths will move the string to and fro by different amounts as they go up and down the string, and the movements from vibrations of different wavelengths eventually cancel each other out. After a fraction of a second, only the vibrations with wavelengths that fit perfectly in the length of the string are causing the string to move.

The wavelengths that fit nicely along the string of a guitar are known as harmonic wavelengths, and the longest wavelength is known as the first harmonic. Every instrument has several harmonic wavelengths because a wave with half the wavelength of the first harmonic will also fit nicely along the string. The same idea applies to other kinds of instruments. For example, in a flute, the harmonic wavelengths fit nicely in the length of the tube, and in a xylophone, the harmonic wavelengths fit nicely in the length of the bar.

There are two kinds of waves: traveling waves and standing waves. Traveling waves move from one place to another; the sound waves coming from the instrument to your ear are a good example of a traveling wave. A standing wave is fixed in place. The vibrations of a guitar string are a good example of a standing wave because, even though parts of the string move from side to side, the string itself is fixed at the ends.

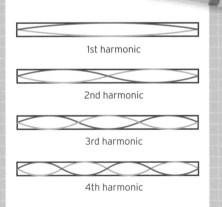

1st harmonic

2nd harmonic

3rd harmonic

4th harmonic

EXPERIMENT: MAKING WAVES

The easiest way to understand what waves are, and how they act, is to make your own. In this experiment you will produce waves of sticky tack balls, water, and sound. While making your waves, think about where else there might be waves around you.

WAVES OF STICKY TACK

In this experiment, you will be making a wave out of sticky tack balls and thread. To make the wave, you will attach the sticky tack balls to some thread and swing the balls from side to side. You should see the peaks of the wave travel along the thread, even though the balls move side to side. Adding some tape will make it easier to see the wave because the tape will be lined at right angles to how the balls move.

YOU WILL NEED:

- 10 pieces of sticky tack (enough to make small balls with a diameter of about 0.5 cm across)
- Piece of thread 60 cm long
- Tape
- Heavy object, such as a book or brick
- Ruler (able to measure in centimeters)

The tape indicates the starting positions of the sticky tack balls.

The balls stop moving for an instant when they are farthest from the tape.

WHAT TO DO:

1. Take your length of thread and tie a knot at one end.

2. Press the pieces of sticky tack onto the thread so that they are roughly spherical, allowing an equal amount of thread between each one. Leave a few centimeters free at each end of the thread. The sticky tack should be pressed tightly enough that the balls don't move along the thread.

3. Tie one end of the thread to a heavy object on a table, or secure it beneath one or more heavy books.

4. Pull the thread taut and stick a piece of tape to the table beneath the full length of the thread. This will mark the thread's position when it is not moving, so that you can compare it to the thread when it is moving.

5. Take the free end of the thread and move it side to side horizontally, at a right angle to the tape. The sticky tack balls will move across the table in the same direction, crossing the line of tape. Vary the speed of the movement and try to find the speed that makes the balls move the most. This should look like the movement of a jump rope.

WHAT HAPPENS?

You should find that when you get the speed of the movement just right, the balls move more quickly, and it is easy to keep this movement going. The balls move the fastest when they cross the line, and the slowest when they are farthest from the tape.

Try it again, but with the balls in different positions, so that the distances between them are all different. Repeat the side-to-side movement of the thread, and see how the balls move. Is it easier or harder to make the balls move around? You should find it harder to keep this kind of movement going.

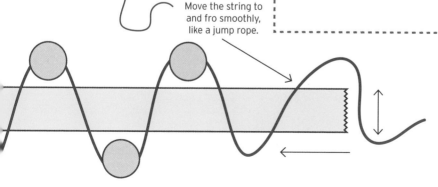

Move the string to and fro smoothly, like a jump rope.

WATER WAVES

In this experiment you will make water waves in a rectangular container by moving the water side to side. This should look like a wave machine in a swimming pool, which makes the surface of the water move up and down.

YOU WILL NEED:

- Water
- Plastic food container, at least 20 cm long, 10 cm wide, and 10 cm deep
- Tray (large enough to hold the plastic container)
- Cardboard
- Scissors
- Food coloring (optional)
- Ruler (able to measure in centimeters)

WHAT TO DO:

1. Place the plastic container on the tray, which will catch any spilled water. Using the scissors, cut out a piece of cardboard that is almost as wide as, and 5 cm taller than, the container.

2. Place the cardboard in the container.

3. Fill the container halfway with water. You can add some food coloring to make the water easier to see. Move the cardboard backward and forward to make small waves. Experiment with the speed of the movement.

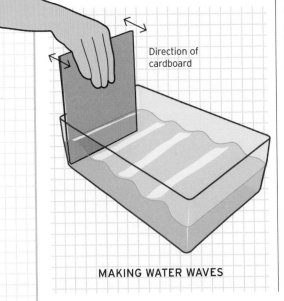

Direction of cardboard

MAKING WATER WAVES

WHAT HAPPENS?

You should find that some speeds are easier than others, and these are the speeds that make the biggest waves.

The water is confined to the size and shape of the container, and waves that fit that shape exactly are the harmonics of the container (see page 47 for more on harmonic wavelengths). It can take some time to find the harmonics, but it becomes quite obvious when you find the right speed.

SOUND WAVES

In this experiment you will be making sound waves by rubbing the rim of a glass with your finger, which makes the glass vibrate and produces a sound wave. Even though you can't see the wave, like you could with the balls and the water, it's clear that there is a wave.

YOU WILL NEED:

- **Small glass with a stem (make sure the sides of the glass aren't too thick)**
- **Water**

WHAT TO DO:

1. Place the glass on a flat surface and hold it down firmly with one hand.

2. Wet the index finger of your other hand and rub it continuously over the rim of the glass.

3. Move your finger smoothly around the rim, maintaining a constant pressure, without pressing too hard or too softly. You should be able to generate a pure note from the glass. It can take some practice, so be patient, and ask an adult for help if you find it difficult. Try making sounds from different shapes and sizes of glass. Try filling the glasses with different amounts of water to see how that affects the sound.

WHAT HAPPENS?

When you rub your finger on the glass, you cause the glass to vibrate. In the case of the guitar (see page 47), the first harmonic is produced because that wavelength fits perfectly on the string. In the case of the glass, its shape and how much of it is free to vibrate are what determine its harmonics. This is why changing the amount of water in the glass changes its pitch—the more water is added, the less space is available, and so the higher the pitch.

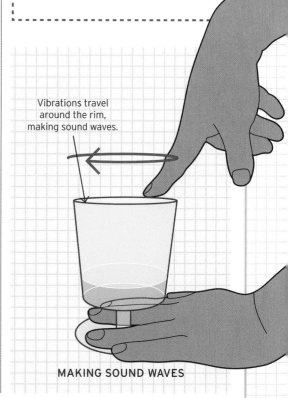

Vibrations travel around the rim, making sound waves.

MAKING SOUND WAVES

DISCOVER: THE ELECTROMAGNETIC SPECTRUM

The light you see around you comprises electromagnetic waves, and visible light is part of what is called the electromagnetic spectrum. Most of the electromagnetic spectrum is invisible to us, but its effects can sometimes be seen and felt under certain circumstances.

THE DISCOVERY OF LIGHT AS A WAVE

Several centuries ago, scientist Sir Isaac Newton placed a glass prism in front of a beam of light. It showed that white light could be separated into different colors, making a rainbow. This happens because light travels slower in glass than in air and because different wavelengths of light have different colors. When light hits a glass prism, it slows down. Speed is related to the frequency and wavelength of light by the following equation:

$$Speed = Frequency \times Wavelength$$

SIR ISAAC NEWTON

The change in the speed is different for different wavelengths in glass, and so the light slows down by different amounts for different wavelengths. When the speed of the light changes, the wavelength of the light also changes to balance the equation. This process is known as refraction, and this is why the light appears to change direction. The different wavelengths changing speeds by different amounts is known as dispersion, and this is why the light passing through the prism spreads out into the rainbow.

BEYOND THE RAINBOW

There is more to light than just the rainbow. If you could see beyond the red end of the rainbow, you would be able to see infrared waves, microwaves, and radio waves. Beyond the violet end of the rainbow, you would be able to see ultraviolet waves, X-rays, and gamma rays. The visible part of the light spectrum is a tiny amount of the whole electromagnetic spectrum. Humans can only see a small range of wavelengths of light.

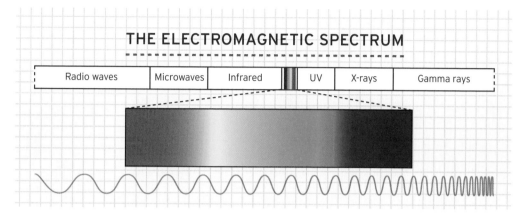

THE ELECTROMAGNETIC SPECTRUM

| Radio waves | Microwaves | Infrared | | UV | X-rays | Gamma rays |

THE REST OF THE SPECTRUM

Electromagnetic waves are incredibly versatile, and without them there would be no radio waves, no infrared sensors for remote control, and no X-ray imaging in hospitals.

Waves on the electromagnetic spectrum interact with matter to create different patterns. For example, when sunlight hits oil, it gets reflected, and different colors are visible on the oil's surface. When a wave meets an obstacle that is about the same size as its wavelength, the wave will change its direction. This is known as diffraction. On a very small scale, the diffraction of X-rays is used to find out about the structure of matter. On a larger scale, radio waves can diffract off mountains, making it difficult to find a good signal.

Electromagnetic waves contain electric and magnetic fields that can move objects with an electric charge, as well as magnetic objects. A good example of this is how Wi-Fi works. Wi-Fi is a kind of electromagnetic wave that is used to send information. A remote control interacts with a TV thanks to electromagnetic waves.

Electromagnetic waves are made up of particles of light called photons. A smaller-frequency photon carries less energy than a larger-frequency photon. That's why infrared radiation (which has a frequency smaller than visible light) feels warm, but ultraviolet light (which has a frequency larger than visible light) can cause sunburn. Generally, higher-frequency electromagnetic waves carry more risk, and lower-frequency electromagnetic waves are not very harmful at all. It's safe to be in a room full of cell phones and radio waves, but it is necessary to limit how much higher-frequency radiation—such as X-ray radiation—people receive.

EXPERIMENT: MAKE A RAINBOW

Following a rainstorm, you might see a rainbow in the sky. You might also see a rainbow when the sun shines through glass or reflects off the surface of oil. This is because different colors of light have different wavelengths, and different wavelengths spread out differently.

Rainbows in the sky are made when the sun shines through water droplets in the air. You can make your own small rainbows with water.

YOU WILL NEED:

- Small glass
- Water
- Sheet of paper
- Powerful flashlight (or a sunny day)

WHAT TO DO:

If you are using a flashlight:

1. Fill the glass three-quarters full with water.

2. Place the glass on the piece of paper and shine the flashlight on it. Move it around until you make a rainbow. You should be able to make a small one; it's easier to see it in a darker room.

WHAT HAPPENS?

When light moves through water, it refracts and separates out the different wavelengths (see pages 52-53 for more on refraction). Each wavelength has its own color, and the colors spread out. When the light shines through the water, it makes rainbows; their shapes depend on the shape of the glass and the angle of the light.

The first rainbow happens when the light reflects off the water once and refracts. To get the second rainbow, the light needs to reflect off the water twice and refract twice. The second reflection means that the order of the rainbow's colors will be reversed compared to the first rainbow. It also means that the second rainbow will be fainter because not all the light will reflect once, and even less light will reflect twice. The sun produces enough light that a second rainbow is usually visible, but a flashlight usually does not produce enough light to make the second rainbow visible.

If you are using sunlight:

1. Fill the glass three-quarters full with water.

2. Place the glass on a table so that half of it is illuminated by the sun. Place the paper on the other side of the glass; you should see a rainbow appear. You may even be able to make a large rainbow on a wall. See if you can identify a secondary rainbow (a fainter second rainbow outside the main rainbow).

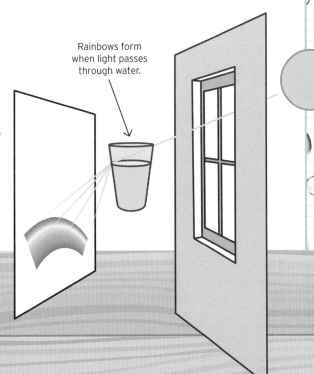

Rainbows form when light passes through water.

LEARN ABOUT: HOW FAST ARE WAVES?

Different waves travel at different speeds; this can be due to a whole range of factors. When a wave travels in a substance, the substance is called the medium.

A wave will usually travel faster in a denser medium because most waves are actually molecules bumping into each other and passing on energy—so the closer the molecules are to each other, the faster the wave will travel.

The fastest anything can travel is at the speed of light in a vacuum. This is because light is a very special kind of wave that does not need a medium in which to travel.

POP QUIZ: RAPID WAVES

There are different examples of waves below. See if you can arrange these waves in order of speed from slowest to fastest:

Sound waves in water: If you go swimming or diving, you might notice that things sound different underwater. Sound waves travel faster underwater, and that can make it sound like things are closer than they really are. Sounds are also distorted compared to normal because water and air carry wavelengths of sound differently.

Water waves in the ocean: In the ocean, all kinds of things can cause waves in the water, including ships and animals moving around. The waves that last the longest are caused by the wind moving across the water. These ocean waves can travel for hundreds of miles.

Light waves: Light can be seen all around us, even from the other side of the visible universe. It doesn't matter how far light travels; whether from your hand to your eye, or from the moon to your eye, the speed of light is a constant.

Waves in earthquakes: An earthquake happens when huge regions of land rub against each other. When this happens, seismic waves are sent through Earth, and they can travel all the way to the other side of the world. The first kind of waves to be detected are called p waves (pressure waves).

Sound waves in the air: Sound waves are special because they can travel through air, water, and even solid objects. Sound waves travel faster through more dense objects (so sound waves travel faster through the metal of pipes compared to through air). Of course, you are most used to hearing sound waves in the air.

LEARN ABOUT: BEYOND VISIBLE LIGHT

Different parts of the electromagnetic spectrum are used for different purposes, depending on the wavelength of the waves. Smaller wavelengths can be used to probe small distances or small objects, and larger wavelengths can be used to interact with larger objects.

POP QUIZ: THE ELECTROMAGNETIC SPECTRUM

Can you connect these different parts of the electromagnetic spectrum with their uses? There might be more than one use for each part of the spectrum. You can find the answers to these questions by asking an adult to help you research online, or searching at your local library.

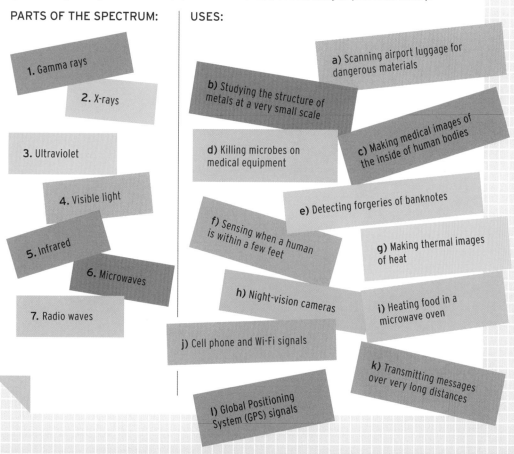

PARTS OF THE SPECTRUM:

1. Gamma rays

2. X-rays

3. Ultraviolet

4. Visible light

5. Infrared

6. Microwaves

7. Radio waves

USES:

a) Scanning airport luggage for dangerous materials

b) Studying the structure of metals at a very small scale

c) Making medical images of the inside of human bodies

d) Killing microbes on medical equipment

e) Detecting forgeries of banknotes

f) Sensing when a human is within a few feet

g) Making thermal images of heat

h) Night-vision cameras

i) Heating food in a microwave oven

j) Cell phone and Wi-Fi signals

k) Transmitting messages over very long distances

l) Global Positioning System (GPS) signals

DISCOVER: HOW MICROWAVE OVENS WORK

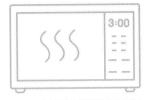

A microwave oven is a good example of how physics can simplify things at home. It is often taken for granted that they heat up food quickly and evenly, but the microwave oven is a relatively modern invention and works very differently from a heat-based conventional oven.

HEATING WATER IN THE MICROWAVE

Microwaves are part of the electromagnetic spectrum, with a wavelength of about 2.5 cm (1 in.). Microwaves are made when a charged object moves backward and forward, and this is done with what is known as an oscillator. The water wave experiment (see page 50) is a good way to visualize how these microwaves are made, with the cardboard performing the role of the oscillator. For both water waves and microwaves, the shapes of the waves are literally defined by the shape of the vessel where they are made. This vessel is called a waveguide because it directs the waves. In this way, microwaves can be made at any size, and when the size of the waveguide is just right for the oscillator, a lot of energy will be transferred to the microwaves.

A microwave oven works by choosing a wavelength that matches water molecules. What does that mean? A water molecule is made up of an oxygen atom and two hydrogen atoms.

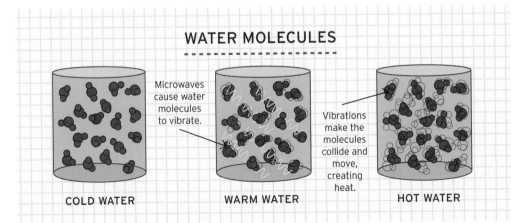

WATER MOLECULES

Microwaves cause water molecules to vibrate.

Vibrations make the molecules collide and move, creating heat.

COLD WATER   **WARM WATER**   **HOT WATER**

The oxygen atom is very slightly negatively charged, whereas the hydrogen atoms are very slightly positively charged. When they feel the effects of the microwaves, they try to align their charges with the microwaves. First the waves move the molecules in one direction, then the opposite direction, and back again, millions of times per second. All this movement heats up the water because heat is a form of kinetic energy. This is how a microwave oven can heat up food quickly and evenly so that it becomes hot all the way through. Compare this to an oven or a toaster, which heats food more slowly from the outside.

IN A BAG OF POPCORN

Kernels covered in hot oil.

Unpopped kernels covered in oil.

COLD METAL SHEET   HOT METAL SHEET

HEATING POPCORN IN THE MICROWAVE

Microwave popcorn uses a special metal lining inside the bag. Metals act like a "sea" of electrons. The atoms are closely packed together, and they are usually arranged in a tight lattice, but the electrons are free to move around. As soon as any electromagnetic wave hits a metal, the electrons will move around, and they can match any wavelength of electromagnetic wave. This means that a lot of heat is produced very quickly (and this is also why it's usually a bad idea to put metal into a microwave oven).

There is also a small amount of oil in the bag of popcorn, and this oil heats up very quickly. Oil does not absorb as much heat as water, so its temperature increases much more quickly than the temperature of water would increase. The heat is then transferred very quickly to the popcorn, and the water inside the kernel is converted to steam. Since the steam is produced in a confined space, the kernel ruptures and pops. Although the oil helps to cook the popcorn, it is not needed to make the kernels pop.

EXPERIMENT: MICROWAVES AND THE SPEED OF LIGHT

Microwaves are a form of electromagnetic radiation, and that means they travel at the speed of light. The speed of light is so fast that it's very difficult to measure. Instead, it's possible to measure it indirectly with a microwave oven.

YOU WILL NEED:

- Microwave oven
- Microwave-safe plate
- Stick of butter or bar of chocolate
- Bag of microwave popcorn
- Scissors

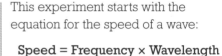

This experiment starts with the equation for the speed of a wave:

Speed = Frequency × Wavelength

The frequency of the microwaves in a microwave oven is 2,450 MHz. As explained on page 58, the oscillator in microwave ovens moves electric charges backward and forward to make the microwaves. The frequency of 2,450 MHz means that the oscillator goes back and forth 2,450,000,000 times per second, making microwaves that also go back and forth at the same rate.

In order to measure the wavelength of the microwaves, you need to make some waves and find the regions where there is the largest transfer of energy. Microwave ovens create hot spots and cold spots. To heat food evenly it must be rotated on a turntable. For this experiment, take out the turntable from the microwave oven.

WHAT TO DO:

1. Unwrap the stick of butter or bar of chocolate and place it on the center of the plate.

2. Place the plate in the center of the microwave oven. Turn the oven on at low power for one minute.

3. See if the butter or chocolate has started to melt, and if not, heat it for another minute. Once the butter or chocolate has started to melt, turn off the microwave oven and remove the plate. (Be careful: it might be hot.)

4. The butter or chocolate will be melted where the hot spots are, and the distance between the hot spots is related to the wavelength of the microwaves. Measure the distance between the centers of two hot spots, and record this distance as d. This is half the wavelength of the microwaves.

5. Multiply the value of d by 2 to get the wavelength, then multiply it by the frequency (2,450 MHz) to get the speed of the microwaves. What speed do you get? You might want to convert it to miles per hour. To do this, divide the speed by 63,360 (inches per mile) and multiply by 3,600 (seconds per hour). A commercial airplane travels at about 550 mph. How does this compare to the speed of microwaves (and therefore of light) that you measured?

WHAT HAPPENS?

Microwave ovens work by making microwaves, which cook food by heating up water. A microwave oven makes waves that fit nicely inside the waveguide, which means there are hot spots and cold spots. Two hot spots are separated by exactly one half wavelength, so the distance between two melted areas of food will be equal to half the wavelength of the microwaves. The turntable in the microwave oven ensures that the food is evenly cooked because all parts of the food move through the hot and cold spots, and everything in between.

By measuring the wavelength and knowing the frequency of the microwaves, it is possible to calculate the speed of the microwaves using the equation for the speed of a wave (see page 52). This speed is the same as the speed of light, which is about a million times faster than an airplane.

Hot spots happen where waves vary the most.

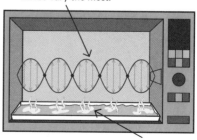

Hot spots melt the chocolate.

DISCOVER: MAKING ELECTRICITY

Electricity is everywhere, and it is used all the time. From alarm clocks to airplanes, it plays a key role in everyday life. Although it is a modern invention, it's hard to imagine a world without it.

HOW POWER PLANTS WORK

When you look at a compass, it points north. The needle of the compass moves until it lines up with Earth's magnetic field. What would happen if you held a magnet close to the compass? The needle of the compass would rotate to face the magnet, of course. What if you kept rotating the magnet around the compass? The needle of the compass would have to keep rotating to point to the magnet. By moving the magnet around, it is possible to keep the needle moving around.

It's not just magnetic objects that react to magnets. Electric charges will also move around when in a magnetic field (see pages 38–39 for a reminder). If you move a magnet up and down next to a metal wire, the electrons in the metal wire will move from side to side as the magnet passes by. This is how a traditional power plant works. Inside the power plant there is a giant spinning coil of wires inside a huge stationary magnet. As the coil of wires spins, the electric charges respond to the effect of the magnet, and they start to move. This produces electrical

WIND POWER

SOLAR POWER

energy, which can be used to power all kinds of machines and gadgets. The part of the power plant that turns the coil of wires is called the dynamo.

KEEPING THE LIGHTS ON

To make electricity from a power plant, all that is needed is to keep the dynamo spinning. That sounds easy enough to do, but it's actually quite a challenge. Wind power plants use the wind to turn giant blades, which turn the dynamo. That sounds like an obvious way to make something spin, but wind power accounts for only about 6.5% of the United States' electrical energy.

Nearly two-thirds of the United States' electrical energy comes from burning coal, oil, and gas. These are called fossil fuels, and there are two major problems with them. First, they create pollution, and second, they will eventually be used up. Fossil fuel power plants burn the fuel to make steam, and the steam can be used to turn the dynamo. About another 1.5% of the United States' electrical energy comes from burning wood and waste.

Another 20% of American electrical energy comes from nuclear power plants. You can read more about nuclear energy in Chapter 4. Nuclear power plants generate electricity by making steam in the same way that fossil fuel plants do, but there is no air

HOOVER DAM

pollution. Instead, they produce nuclear waste, which must be stored safely.

Big dams like the Hoover Dam use falling water to turn dynamos. These kinds of dams account for 7% of the United States' electrical energy, and they have an obvious impact on the local environment.

Finally, solar power uses special solar panels made of solar cells to create electrical energy without dynamos. These work in a similar way to how the leaves of a plant work. When sunlight falls on a surface, it transfers some energy. This can be felt on a sunny day as heat. A plant converts this to energy using photosynthesis, and a solar cell converts this to energy by moving charged particles.

LEARN ABOUT: CIRCUITS

Nearly all gadgets require an electricity supply. This might be in the form of a battery or a power outlet. Either way, the method is the same, with wires transmitting different amounts of electrical energy. By connecting these wires into circuits, it is possible to extract the energy to do useful work, for example, to power a toaster.

POP QUIZ: ELECTRICAL CIRCUITS

A microwave oven takes electrons with a lot of energy, extracts this energy, and then returns the low-energy electrons to a power outlet. As it does this, the electrons move across the potential difference (or voltage) of 110 V. The potential difference is a bit like height, and moving across a large potential difference is like falling off a cliff—it releases a lot of energy.

The number of electrons moving per second is described by the current. Given that the potential difference for 110-V outlets is always the same, the rate of energy transferred will depend on the current. The rate of energy transfer, called the power, is given by the equation:

Power = Current × Potential difference

The current and potential difference are related by what is called the resistance:

Potential Difference = Current × Resistance

The unit of power is called watt, the unit of current is ampere, the unit of potential difference is volt, and the unit of resistance is ohm.

The microwave oven can be represented with an electric circuit:

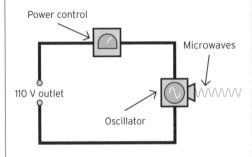

If the power rating of the microwave oven is 800 W, and the potential difference is 110 V, what is the current traveling through the microwave oven? What is the resistance of the microwave oven?

Suppose you want to defrost some meat very slowly. This means you would need to lower the microwave oven's power. Let's say you want the power to decrease by a factor of three. How would you change the resistance to achieve this?

LEARN ABOUT: DIFFERENT POWER PLANTS

Everyone uses electricity, even if they don't think about it. Every time you watch television you are using electricity. Every time you walk down the street at night, you pass streetlights that are powered by electricity. Where does this electricity come from?

RENEWABLE AND NONRENEWABLE POWER

There are many ways to generate electricity. One way is to burn fuel to make steam to turn a turbine—a large device with fins, which works together with a dynamo to produce electricity. Some ways of making electricity can be used forever because they are renewable. If a way of making electricity has limited sources, it's known as nonrenewable, and it will eventually run out.

Generally, renewable energy sources generate less electrical energy than nonrenewable energy sources, but that is slowly changing as technology improves. Nonrenewable energy sources cause large amounts of pollution, or waste, which must be cleaned up. Burning fuel also leads to the release of carbon dioxide, which is causing climate change. In order to continue using electrical energy in the future without damaging the environment, it will be necessary to take it from renewable sources.

At the time of writing, renewable energy sources account for just over one-sixth of all electrical energy in the United States. The technology for renewable energy sources must improve in order to increase that proportion.

(NON)RENEWABLE?

Can you divide the following sources of electrical energy into renewable and nonrenewable?

1. Coal-burning power plant
2. Solar power plant using solar cells
3. Wind power plant
4. Oil-burning power plant
5. Hydroelectric (falling water) power plant
6. Gas-burning power plant
7. Solid waste-burning power plant
8. Tidal power plant (using tides in the ocean)
9. Nuclear power plant
10. Geothermal power plant (using Earth's heat)

EXPERIMENT: MAKING A BATTERY MOTOR

Electricity can be useful for making lights work and computers run, but how can electrical energy be turned into motion? The answer is quite simple—the process of electrical energy generation is reversed. In this experiment, you will make a small, simple motor.

YOU WILL NEED:

- AA or AAA battery
- Metal screw
- Small neodymium disk magnet
- Wire (around 7.5 cm [3 in.])
- Sticky tape (optional)

ADULT SUPERVISION REQUIRED

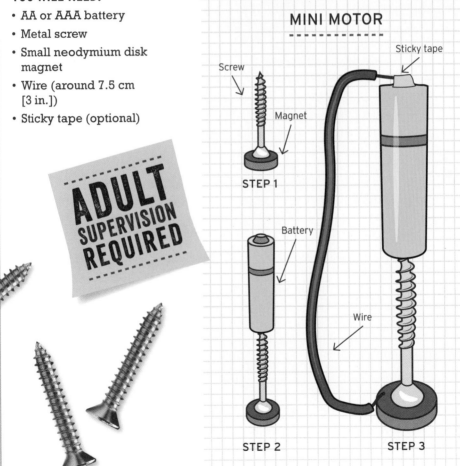

MINI MOTOR

Screw

Magnet

STEP 1

Sticky tape

Battery

Wire

STEP 2

STEP 3

MOTORS AND DYNAMOS

Dynamos work by spinning a coil of wires near a magnet. Electric motors work by reversing this process. If you have a source of electricity and a magnet, you can combine them to make something spin. This is because it's all the same physics at work. If you can turn a coil near a magnet to make electricity, you can use electricity near a magnet to turn a coil.

WHAT TO DO:

1. Place the flat head of the screw on the magnet.

2. Hold the battery up and attach the point of the screw to the negative end of the battery. It should stay in place because of the magnet.

3. Hold one end of the wire to the positive end of the battery (you can use tape to keep it in place). Gently touch the other end of the wire to the magnet and see what happens.

4. Try turning the magnet upside down and see what happens. Then turn the battery upside down and see how that changes things.

WHAT HAPPENS?

When you connect the wire to the battery and the magnet, you make an electrical circuit. This makes the electric charges flow through the magnet and the screw. Spinning a coil next to a magnet makes electricity flow, so if you do the reverse and make electricity flow near a magnet, you should cause the coil to spin. In this case the coil is the screw.

When you turn the magnet upside down, you turn the magnetic field upside down, and this causes the screw to turn in the other direction. In the same way, turning the battery upside down reverses the direction of the flow of electricity, so the screw turns the other way.

ELECTRIC MOTORS

Electric motors come in all sizes and are used around the world for all kinds of purposes. Common examples include electric toothbrushes, fans, and food mixers. The largest electric motors in the world are used in big ships and industrial pipes. Whether an electric motor is being used to push a ship or to beat some eggs, the principle is the same as the tiny motor you made. An electric circuit, combined with a magnet, can make objects turn, and this can be used to make many kinds of useful machines.

DISCOVER: WHAT IS LIGHT?

What is light made of? It might sound like a simple question, but it is one that has puzzled scientists for hundreds of years. This is because light acts in unexpected ways: sometimes it acts like it is made of waves, and sometimes it acts like it is made of particles.

LIGHT BEHAVES LIKE A WAVE

On a rainy day, it is sometimes possible to see a rainbow. That's because light from the sun hits droplets of water in the atmosphere and is separated out into its different colors, forming circular bands in the sky. This is what you would expect if light were a wave; the different colors spread out because they have different wavelengths.

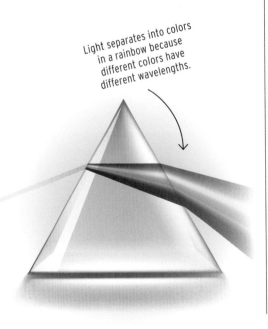

Light separates into colors in a rainbow because different colors have different wavelengths.

LIGHT BEHAVES LIKE A PARTICLE

People who spend a lot of time outdoors in the summer should use sunscreen to protect their skin. This is because the sun gives off ultraviolet light, which can cause sunburn. If someone spent all day indoors, where there is no ultraviolet light, they would not need to use sunscreen, and it would not matter how much light fell on their skin. Why are these people much more likely to get sunburned after 30 minutes sitting in the sun than from 12 hours sitting under a glowing light bulb?

This is because ultraviolet light from the sun has a shorter wavelength than light from a light bulb, and a shorter wavelength means it has more energy. Light is made of particles called photons, and it is the amount of energy per photon that causes sunburn. A single high-energy (ultraviolet) photon can hit your skin, transfer all its energy to the atoms in your skin like a bullet, and cause a burn. A single low-energy (visible light) photon can hit your skin and bounce off harmlessly. When looked at in this way, light seems to behave as though it is made of particles.

LIGHT AS WAVES AND PARTICLES

How can it be that light sometimes behaves like it is made out of waves, and sometimes as though it is made out of particles? The easiest way to understand it is to think of light being made of photons that can act together as waves, in the same way that water molecules make up water. A photon has a wavelength that humans see as color, and its wave looks like a mini wavelet. As the photon moves forward, the wavelet moves from side to side. When a very large number of photons move together, they can behave like a wave, for example, in a laser beam.

This is not the end of the story. Scientists have studied single photons to see how they act, expecting them to behave like particles. In fact, a single photon will still act like a wave sometimes.

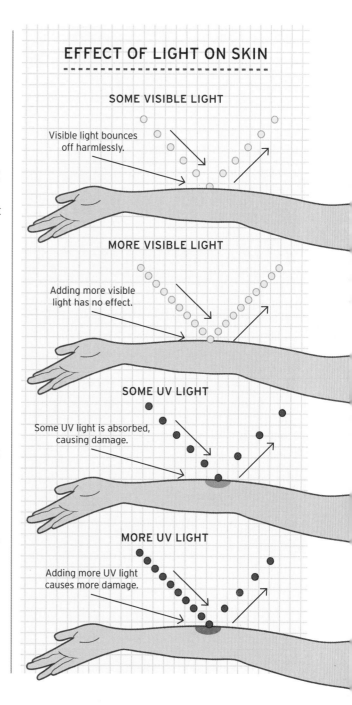

EFFECT OF LIGHT ON SKIN

SOME VISIBLE LIGHT

Visible light bounces off harmlessly.

MORE VISIBLE LIGHT

Adding more visible light has no effect.

SOME UV LIGHT

Some UV light is absorbed, causing damage.

MORE UV LIGHT

Adding more UV light causes more damage.

EXPERIMENT: FILTERING LIGHT

If you see a 3-D movie at a movie theater, you will be given special glasses. They make use of a very simple, strange fact about how light behaves to filter light in a way that makes 3-D movies possible.

POLARIZED LIGHT IN MOVIE THEATERS

When you want to drop an envelope in a mailbox, you need to line the envelope up with the mail slot. If the slot is horizontal, you hold the envelope horizontally. Photons of light act in a similar way when they pass through filters. If the wavelet of the photon is lined up with the filter, the photon will pass through. If the wavelet of the photon is at 90 degrees to the filter, the photon will not pass through. For angles in between, the photon might or might not pass through the filter. If the wavelet of the photon is at 45 degrees to the filter, there is a 50% chance it will pass through. If the photon passes through the filter, it will always be lined up with the filter afterward.

ENVELOPES AND PHOTONS

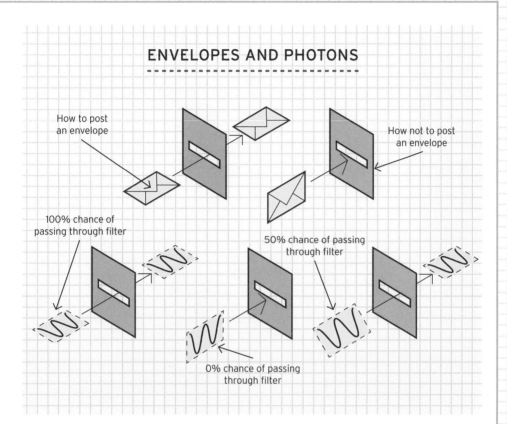

How to post an envelope

How not to post an envelope

100% chance of passing through filter

50% chance of passing through filter

0% chance of passing through filter

MOVIE MAGIC

Movie theaters that show 3-D movies use glasses with special filters (the illustration above shows how the filters work). These filters control what light is seen by the left eye and right eye of the viewer. In this way, the 3-D movie glasses fool your brain into thinking you are looking at solid objects instead of at a flat screen. Your left and right eyes see things from slightly different positions, and that allows you to judge how far away things are. This is known as depth perception (see the box on page 73 for more on this). So if a 3-D movie theater wants to make you think you are looking at solid objects, it just needs to make your left eye see one thing, and your right eye see another. If it does this just right, it will make you feel as if you are really there, in the scenes of the movie. One simple way to do this is to filter the light that reaches your left eye and the light that reaches your right eye.

In this experiment, you will use light filters to control light. By placing a filter in a certain way, it is possible to allow some wavelets of light through the filter, but stop other wavelets.

YOU WILL NEED:

- **Polarizing film (for example, from disposable 3-D movie glasses)**
- **Scissors**
- **Sticky tape**
- **Sheet of white paper**

WHAT TO DO:

1. To keep track of orientation, place tape along one edge of the polarizing film. Cut the film into three strips, so that each piece has some tape.

2. Put one piece of film on the paper. The paper should look darker under the film because only some light can get through. Take a second piece of film and turn it over. Place it over the first piece of film and see how dark the paper looks under both pieces of film. It should be just as dark as before. Slowly rotate the second piece of film

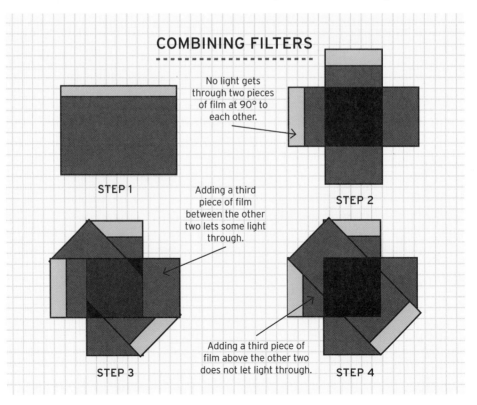

COMBINING FILTERS

No light gets through two pieces of film at 90° to each other.

STEP 1

STEP 2

Adding a third piece of film between the other two lets some light through.

STEP 3

Adding a third piece of film above the other two does not let light through.

STEP 4

until it is at 90 degrees to the first piece of film. You should find that no light gets through when the two pieces of film are at 90 degrees to each other.

3. Slide the third piece of film between the other two. Rotate it slowly and see how different amounts of light get through all three pieces of film. It seems that adding a third layer can actually increase the amount of light that can get through.

4. Try putting the third piece of film on top of the other two pieces of film. You should find that, once again, no light gets through. It is only by placing the third piece of film between the other two that light can get through.

WHAT HAPPENS?

When light reaches the piece of film, only half of it gets through because not all the photons line up with the filter. Adding a second piece of film at 90 degrees stops all light because the photons were forced to line up with the filter in the first piece of film, so they could not line up with the filter in the second piece of film.

If you add the middle filter at 45 degrees, it's possible to let some light through from the top filter to the middle filter, and from the middle filter to the bottom filter. Adding a filter can actually increase the amount of light that reaches the paper.

DEPTH PERCEPTION

Try this: hold your thumb out in front of you at arm's length. Focus on your thumb and then close one eye at a time while still looking at your thumb. There shouldn't be much difference in the view from each eye. Now, move your thumb toward you so that it is 7 cm (about 3 in.) from your nose. Look at your thumb, first with your left eye closed, and then with your right eye closed. This time each eye can see a different side of the thumb.

The different views you get from your eyes help you to judge how far away things are. This is called depth perception.

7 cm

LEARN ABOUT: SENDING SECRET CODES WITH PHOTONS

In order to send a secret message to someone, a way is needed to scramble and unscramble it—to encrypt and decipher it. The science of encrypting and deciphering messages is known as cryptography, and photons can be used by cryptographers to send secret messages.

POP QUIZ: SECRET MESSAGES

A common way to send messages is to use Morse code. In Morse code, every letter gets converted into dots and dashes. So as long as you can agree on how to send dots and dashes, you can send messages using Morse code.

A •−	H ••••	O −−−	V •••−
B −•••	I ••	P •−−•	W •−−
C −•−•	J •−−−	Q −−•−	X −••−
D −••	K −•−	R •−•	Y −•−−
E •	L •−••	S •••	Z −−••
F ••−•	M −−	T −	
G −−•	N −•	U ••−	

Alice wants to send the following message to Bob: "LETS GO TO THE MOVIE THEATER." She converts this to dots and dashes using Morse code and sends the following:

•−•• • − ••• / −−• −−− / − −−− / − •••• • / −− −−− ••− •• • / − •••• • •− − • −•−

1. Bob receives the Morse code and translates it back to read the message. He sends the following message to Alice. What does it say?

−−− −• • −•− / •−•• • − ••• / −−• −−− / −−− −• / − ••− • •••• −•− •− − •−

2. Bob is worried that anyone can read this message. His friend Eve likes to read his secret messages, and if she could read the dots and dashes, she would be able to find out what he is saying to Alice.

Alice comes up with an idea to use photons. She will send Bob a photon lined up vertically for a dot, and horizontally for a dash. Bob can then use a horizontal filter to read the Morse code. If the photon passes through his filter, he knows it was horizontal and he writes down a dash. If it doesn't pass through his filter, he knows it was vertical and he writes down a dot.

Alice tests the new method by sending her name:

ALICE → .- .-.. .. -.-. . → VH VHVV VV HVHV V

Bob receives and decodes the message, and sends a message back:

VVVV VV / VH VHVV VV HVHV V / H VVVV VV VVV / VV VVV / HVVV HHH HVVV

What message did Bob send to Alice?

3. Bob is still worried that Eve can read the messages because all she needs to do is place her own filter between Alice's and Bob's messages, and neither Alice nor Bob would ever know.

After thinking for a while, Alice comes up with another idea. She knows that if Bob has his filter lined up horizontally and Alice sends a photon at 45 degrees, there is a 50% chance the photon will pass through his filter, so there is a 50% chance Bob will get a dot and a 50% chance Bob will get a dash. However, if Bob also rotates his filter by 45 degrees, the photon will be lined up properly, and he will read the message correctly.

They both rotate their filters and photons, and Alice sends a message to Bob:

VVVV V VHVV VHVV HHH

What did Alice send to Bob?

Bob decides to send the same message back, but this time Eve intercepts the message. Half the photons get through Eve's filter, but Eve cannot read the message because there is no way for her to know which photons were horizontal and which ones were vertical. The photons that pass through Eve's filter are at 45 degrees to Alice's filter, and only half of these photons pass through Alice's filter. That means that Alice only receives one-quarter of the photons Bob sent, and not half, as expected. Now Alice knows that Eve is trying to read the messages.

CHAPTER 3
FORCES AND GRAVITY

DISCOVER...

LEARN...

EXPERIMENT...

DISCOVER: FORCES

Forces allow you to interact with the world around you. Every time you lift an object, or even just walk around, you apply a force in order to do something. When you lift a book off a table, you pull it up, and when you walk, you push against the ground.

NEWTON'S LAWS

In the seventeenth century, Sir Isaac Newton came up with three laws to describe how forces make objects move. These laws can be summarized as:

• **First law:** A stationary object won't move unless a force acts on it. A moving object won't change its speed or direction unless a force acts on it.
• **Second law**: If a force acts on an object, it will be accelerated in the direction of the force.
• **Third law:** Every force is balanced by an equal and opposite force.

For example, let's say that you hit a stationary hockey puck. Before you hit the puck, there is no force acting on it, so it stays still (first law). When you hit the puck with your hockey stick, you apply a force, so it starts to move (second law). When you hit the puck, you feel some resistance as the puck pushes back on the stick (third law). If the puck was already moving before you hit it, it would have moved in a straight line at a constant speed until

you hit it (first law). Newton's laws provide a very simple way to understand how forces act.

The second law states how forces make objects accelerate, and this is given by the equation:

Force = Mass × Acceleration

This means that if you double the force, you double the acceleration. Think about how a hockey puck moves faster if you hit it harder. It also means that heavy objects won't accelerate as much when acted on by the same force. That's why it takes more effort to throw a baseball than a table tennis ball. If you want them to both go at the same speed, you need to throw the baseball a lot harder to give it a larger force.

CONTACT FORCES AND LONG-RANGE FORCES

Most forces you come across are called contact forces. These are pushes and pulls that take place when objects touch each other. When you climb the stairs, your feet push against the steps, and you might pull yourself up using the banister. If two objects aren't touching each other, then there is a "long-range" force. Gravity is a good example of a long-range force, and so is electromagnetism. You have seen the electrostatic and magnetic forces working over a few inches, so they are not contact forces.

Forces act at all different scales. There are forces inside atoms that make up everything around us, keeping objects together. There are forces that make cars move. At the largest scale, there are forces that keep Earth orbiting the sun.

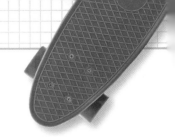

EXPERIMENT: TESTING NEWTON'S THIRD LAW

According to Newton, every action has an equal and opposite reaction. When you use a skateboard, you push against the ground, and it pushes back on you with an equal force. If you pushed on something with the same mass as you, you might cause that to move as well.

YOU WILL NEED:

- 2 skateboards
- Helmets, knee pads, elbow pads
- Friend who is roughly the same weight as you
- Long pole (a mop or broom handle will do)
- Rope

WHAT TO DO:

For this experiment, you will be standing or sitting on skateboards, so make sure to wear a helmet, knee pads, and elbow pads in case you fall off. Find a flat area of concrete.

1. First, get your friend to stand on the ground, while you stand or sit on the skateboard. You and your friend should each grab one end of the pole. Push gently against your friend, and you should move away from your friend. This should not be surprising because this is just like pushing against the ground.

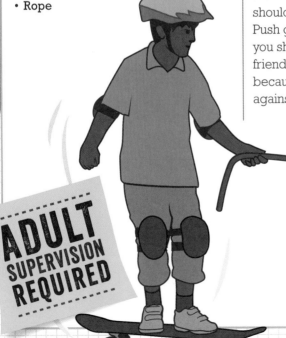

ADULT SUPERVISION REQUIRED

STEP 4

2. Repeat the experiment but instead have your friend gently push on the pole. You should move away in the same way as you did before. This is because it doesn't matter whether you push on your friend, or your friend pushes on you. As long as you both hold the pole, you will both feel a force, and you will move away from your friend.

3. For the next part, ask your friend to get on the other skateboard (remember the helmet and pads). You should both hold the pole (one at each end) and both push gently. What happens? Next, take turns pushing.

4. Swap the pole for the rope, and this time try pulling on the rope. Do you both move, or does one of you move? What would happen if you had three people in a line, connected by pieces of rope, and the person at the center pulled on both ropes?

5. Finally, ask an adult to join in. You should be on one skateboard, and the adult should be on another. When you push gently against the adult using the pole, do you move faster or slower than the adult? What if the adult pushes against you? See if you can get two adults to push against each other. Do they travel faster or slower than when you try with a friend?

WHAT HAPPENS?

Whenever you push or pull on something, you apply a force. That force must be balanced by an equal force pushing or pulling back. When you push against your friend, whether or not they are on a skateboard, you feel a force push you back that makes you move. When you are both on skateboards, you should move apart with about the same speed as each other. That's because you both experience the same force, and you have roughly the same mass, so you have roughly the same acceleration. When you push against an adult on a skateboard, you probably move faster than they do because you both experience the same force, and they have more mass than you, so they have a smaller acceleration.

LEARN ABOUT: SPEED GRAPHS

Speed describes how quickly something is moving, and acceleration states how quickly its speed is changing. If you know the speed of a car throughout its journey, it should be possible for you to work out how far it travels. Speed graphs allow you to do just that.

DISTANCE, SPEED, AND ACCELERATION

Distance and speed are related by time in the following way:

Distance = Speed × Time

It's easy to see that if you travel at 30 m/s, you will travel twice as far in the same time than if you travel at 15 m/s. Acceleration describes how speed changes over time in the same way:

Change in speed = Acceleration × Time

1. Suppose the driver of a car is traveling at 22 m/s. They see a speed limit change up ahead to 13 m/s, and they have about 10 seconds to reduce their speed. What should their acceleration be? What is the average speed of the car as it slows down?

HOW SPEED GRAPHS WORK

Speed graphs work by showing the speed of an object as a function of time. The speed is on the y-axis and the time is on the x-axis. If a bus travels at 10 m/s for 20 seconds, its speed graph would look like this:

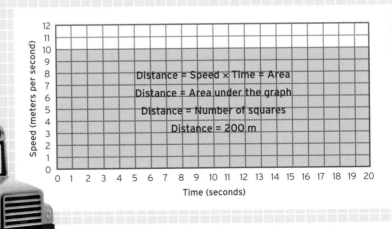

Distance = Speed × Time = Area
Distance = Area under the graph
Distance = Number of squares
Distance = 200 m

The distance traveled is the area under the graph. You can find this by multiplying the speed by the time, or by counting squares. Whichever method you use, you should get the same area, which is 200 m.

2. If a bus accelerates at a constant rate, the line on the graph will create a triangle shape. In the graph on the right, the bus accelerates up to 10 meters per second over 6 seconds. What is the acceleration? You can either measure the slope of the graph, or you can divide the change in speed by the time taken. What distance does the bus travel?

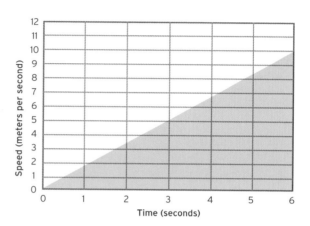

3. The graph on the left shows the journeys of a car (in blue) and a truck (in red). The car travels faster, but takes a break in the middle of the journey. What is the initial acceleration of the car and the truck? Which travels the farthest overall?

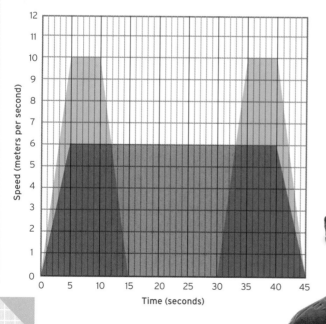

DISCOVER: GRAVITY

There's one force that connects everything in the universe. Everything with mass feels this force, and its effects can be felt from billions of miles away. This is the force of gravity, and on Earth you feel it pulling you down to the surface. It also keeps the moon circling Earth.

Jumping against gravity

GRAVITY ON EARTH

You can feel the pull of gravity on you all the time. Any object with mass will pull any other object with mass, and the more mass an object has, the stronger its gravitational pull. To find out your weight in a certain place, you can multiply your mass by the local gravity. Gravity is nearly the same all over the world, but it's slightly weaker at the tops of mountains. This is because the gravitational force between two objects gets smaller as the distance between them gets bigger. Even so, gravity at the top of the world's tallest mountain (Mount Everest) is only 0.2% smaller than it is at sea level.

The weight of an object is related to its mass by:

Weight = Mass × Acceleration of gravity

The acceleration of gravity is what causes things to fall and to gain speed as they fall. It costs energy when you lift an object against gravity. That's why going up stairs is more tiring than going down stairs. The change in gravitational energy is:

Change in gravitational energy = Weight × Change in height

When you climb the stairs, your muscles have to do a lot more work to give you the extra gravitational energy. When an object falls, it loses its gravitational energy, and that's what makes it speed up.

GRAVITY IN SPACE

Gravity doesn't just pull you down to Earth, it also keeps the moon orbiting around Earth in a circle. Gravity keeps Earth orbiting around the sun as well. The gravity of the moon and the sun even affects the oceans. The high and low tides on Earth are due to the gravitational pull of the moon and the sun. The spring tides are very special high tides that happen when the moon and the sun line up and their combined gravitational pull is at its strongest. The force of gravity between the moon and Earth is so strong that the moon is permanently deformed into an egg shape, and the same side always faces Earth.

On an even larger scale, gravity keeps the galaxy together. This is because gravity is a long-distance force, and it can be felt billions of miles away. Gravity is always attractive, and that means the gravitational pull of extremely heavy objects, like the sun, can be felt over very large distances. Compare this to the electrostatic force between electric charges. Even though the electrostatic charge is much stronger, the positive and negative charges nearly always cancel each other out. That means that you don't often see the electrostatic force in action, and when you do, it tends to be very dramatic (for example, in lightning storms).

Gravity is felt everywhere, even in space.

EXPERIMENT: FALLING UNDER GRAVITY

Gravity is the force of attraction that mass has on other mass. Earth's mass is what creates the gravitational force that pulls everything on Earth down. When you drop something, it falls down, slowly at first, but then gains speed. How quickly does it gain speed when falling?

YOU WILL NEED:

- Unpopped popcorn kernels
- Ruler (able to measure in centimeters)
- Stopwatch
- Sticky notes

WHAT TO DO:

1. Measure some heights on the wall, from the floor to the ceiling. Place a sticky note at 60 cm, 90 cm, 120 cm, 150 cm, and 180 cm.

2. Next, take an unpopped kernel and drop it from the 120-cm marker. Time how long it takes to hit the floor and write this down. A good way to do this is to have the stopwatch in one hand and the kernel in the other. As you release the kernel with one hand, start the stopwatch. Then stop the stopwatch when you hear the kernel hit the floor. You will not have much time, so don't think too hard about it. You may want to practice a few times.

3. Repeat step 2 for all the other height markers, so that you know how long it takes for a kernel to fall 60 cm, 90 cm, 120 cm, 150 cm, and 180 cm. Fill out the first three columns in the table at the top of the next page.

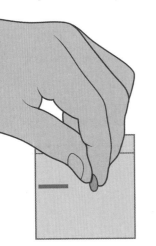

HEIGHT	Time taken (seconds)	Time taken × Time taken	Gravity = 2 × height ÷ Time²	2 × height ÷ average time²
60 cm				
90 cm				
120 cm				
150 cm				
180 cm				

In the second column you will need to multiply time by itself to get the square of the time. In the third column, multiply the height by 2, and divide by the square of the time. What do you notice about the values in the third column? They should be roughly the same value. If you find that they are not very close to each other, you can run the experiment again.

For example, if it took 0.5 seconds to fall 120 cm, the values would look like this:

Time = 0.5 seconds
Square of time = 0.5 × 0.5 = 0.25 s²

Gravity = 2 × Height ÷ Time²

Gravity = 2 × 120 cm ÷ 0.25 s² = 240 ÷ 0.25 = 960 cm s²

4. Repeat the experiment so that you get 10 different times for each height. Add these times up for each height, then divide by 10 to get the average time taken. Now use these times to fill out the fourth column and see if the results look more similar to each other.

WHAT HAPPENS?

You should find that you get the same value for gravity, no matter what height you dropped the kernel from. The acceleration of gravity is the same for everything, so heavy objects and light objects fall at the same rate. If you dropped a basketball and a baseball from the same height and at the same time, they would hit the floor at the same time. This was discovered 400 years ago by Italian physicist Galileo Galilei, and his discovery led to the start of physics as a science.

LEARN ABOUT: GRAVITY AND THOUGHT EXPERIMENTS

Two thousand years ago, ancient Greek scientists believed that heavier objects would fall faster than lighter objects. Nowadays, scientists always test their ideas with real evidence—but it turns out you don't even need experiments to prove that the ancient Greek scientists were wrong about gravity.

A THOUGHT EXPERIMENT

A thought experiment is an idea that you test without doing any practical experimenting. It's a way to help you to understand your ideas and make sure that they make sense. If you can create a thought experiment for an idea that leads to a contradiction, then there must be something wrong with the idea. The ancient

Greek philosopher Aristotle thought that heavier objects fell with a larger acceleration than lighter objects. Two thousand years later, an Italian scientist named Galileo tested this idea and found that all objects fall with the same acceleration.

Imagine that Galileo and Aristotle meet and talk about their ideas. Galileo might talk about a thought experiment using Aristotle's ideas. Galileo would suggest that they think about dropping a heavy object and a light object from a high cliff. According to Aristotle, which object would hit the ground first? According to Galileo, which object would hit the ground first?

ARISTOTLE

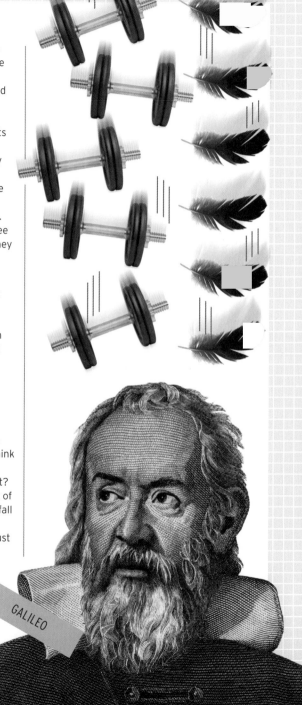

Then Galileo comes up with an idea to make the thought experiment more interesting. He suggests connecting the two objects with a piece of string. This string is very strong, and the only effect it has is that when it is taut, it has tension that pulls on the objects. "That's fine!" says Aristotle. "The two objects will fall until the string becomes taut, then the tension in the string will make the heavy object fall more slowly, and the light object fall more quickly." Galileo asks, "So when the string is taut, will they fall with the same acceleration?" "Of course!" replies Aristotle. "At least we can agree on that. We both agree they fall with the same acceleration when they are tied together."

However, Galileo is not finished with his thought experiment. "So is the heavy object falling more slowly than it was before it was tied to the light object?" he asks. "Yes. The lighter object has the effect of slowing down the heavier object." "Aha!" replies Galileo. "I know how to prove your idea is wrong! I can make the lighter object have the effect of speeding up the heavier object!"

Aristotle thinks that the lighter object will cause the heavier object to fall more slowly. What very small change can Galileo make to the thought experiment to make Aristotle think that the heavy object falls more quickly because of the presence of the lighter object? If Aristotle's ideas make him think the effect of the light object is to make the heavy object fall both more quickly and more slowly, this is a contradiction and his ideas about gravity must be wrong.

GALILEO

DISCOVER: MOMENTUM

Imagine you are riding a bicycle down the street. If you suddenly stop pedaling, what will happen? You don't stop; you continue to move. If you are on a smooth surface, you will carry on with the same speed. That's because of momentum, and momentum is what keeps things moving.

WHAT IS MOMENTUM?

When an object moves, it has momentum. The faster something is traveling, the more momentum it has, and the more mass an object has, the more momentum it has. It takes more effort to stop a bicycle when you are traveling quickly compared to when you are traveling slowly because you have more momentum. It also takes more effort to stop a bicycle if you are wearing a heavy backpack because you will have more mass. The amount of momentum an object has is:

Momentum = Mass × Speed

If you and a friend are on a tandem bicycle, you will have about twice the momentum as just you being on a normal bicycle at the same speed. If you ride at 5 m/s (11.2 mph), you will have twice as much momentum compared to riding at 2.5 m/s (5.6 mph). If you want to change your momentum, you need to change your speed, and that means you have to apply a force. The bigger the change in momentum, the bigger the force that's needed.

The faster you ride your bicycle, the harder it is to stop.

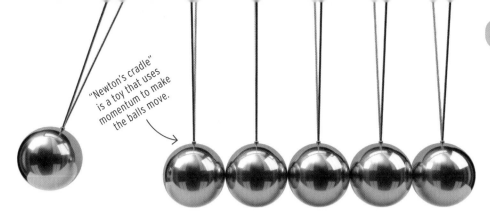

"Newton's cradle" is a toy that uses momentum to make the balls move.

CONSERVING MOMENTUM

When you are ice-skating at a high speed, it can be difficult to slow yourself down because you have so much momentum. A lot of beginners simply stop themselves by crashing into the wall in a controlled way. Changing momentum means applying a force, so what happens if there are no forces, or if all the forces are canceled out? In that case, the momentum will not change, and things will keep moving at the same speed. This is called conservation of momentum.

In the case where two objects collide, conservation of momentum can be written as:

Momentum before collision = Momentum after collision

Imagine you are playing pool, and you hit the cue ball. As it moves, it hits the eight ball, and the eight ball starts to move. If the cue ball hits the eight ball head-on, it will lose all its momentum to the eight ball. This is because the cue ball applies a force to the eight ball, and that force is enough to give the eight ball all the momentum. If the eight ball takes up all the momentum, and the overall momentum stays the same, the cue ball's momentum must be zero after they hit. If the cue ball hits the eight ball at an angle, the eight ball and cue ball will each move off at an angle. But overall, the total momentum will stay the same.

WHAT NEWTON'S THIRD LAW MEANS

Newton's Third Law says that every force has an equal and opposite force. Forces change momentum, so if every force has an equal and opposite force, then every change in momentum must have an equal and opposite change in momentum.

EXPERIMENT: CONSERVING MOMENTUM

Pool balls all have about the same mass, and this makes it easy to predict where they will move after they have hit each other. This means it's not too hard to bounce balls off cushions (the sides of a pool table) and perform trick shots. But what about when things with different masses hit each other? You can use toy cars to find out more about this.

YOU WILL NEED:

- 2 long, smooth planks of wood
- Stack of books, about 30 cm (1 ft.) high
- Several toy cars of different masses
- Tape measure
- Stopwatch
- Kitchen scale

WHAT TO DO:

Even when momentum is conserved, it's sometimes hard to see that this is the case.

1. Set up a racecourse by placing one plank of wood flat on the floor. Make a ramp by placing the end of the second plank of wood on the stack of books and the other end touching the first plank of wood. There should be no gap between the two planks of wood.

2. Label each car with a number, weigh each of the cars, and write down the values. (You can refer to the cars by color or shape if you don't want to physically mark them.)

3. Place a toy car at the top of the ramp and release it. It should travel down the ramp and along the racecourse. Time

how long it takes the car to travel along the ramp and racecourse. Do this for each car and write down how long each car takes. Do they take roughly the same amount of time?

4. Now take two cars that have roughly the same mass. Place one car where the ramp meets the racecourse. Release the other car from the top of the ramp. The car on the ramp should hit the car on the racecourse, and the car on the racecourse will start to move. Time how long it takes between

releasing the car at the top of the ramp and the car at the bottom reaching the end of the racecourse.

How does this compare to the time taken when you had only one car? Try this experiment again with the heaviest car at the top of the ramp and the lightest car at the bottom. Try again with the lightest car at the top of the ramp and the heaviest car at the bottom.

WHAT HAPPENS?

When the cars move down the ramp, they change gravitational energy (see page 84) into kinetic energy. Since the gravitational energy and the kinetic energy both depend on the mass of the cars, each car should reach the bottom of the ramp with the same speed. Let's call this speed VRamp. Suppose the mass of the car on the ramp is MRamp, and the mass of the car on the racecourse is MTrack. Newton's Third Law states that momentum is conserved, so if this is the case, it can be stated:

Momentum before = Momentum after

If the first car loses all its momentum in the collision between the cars, then the second car must take all the momentum.

The momentum is given by the equation:

Momentum = Mass × Speed

That means that the equation can be written:

MRamp × VRamp = MTrack × VTrack

This can be rearranged to express the speed of the car on the track in terms of the masses of both cars and the speed of the car on the ramp (which should be the same for cars of all masses):

VTrack = VRamp × (MRamp ÷ MTrack)

Does this match the results of your experiment? Did the lightest car travel the fastest when it was on the racecourse? Did the heaviest car travel the slowest? Perhaps the heaviest car didn't even make it to the end of the racecourse.

LEARN ABOUT: PLAYING POOL

If you've ever seen good pool players in action, you've probably noticed them sinking the balls as if it's second nature—after playing for so many years, they get a good feel for how to take each shot. But pool is actually just about physics.

STRAIGHT SHOTS

Sinking a ball when everything is lined up is easy. What you want to do is hit the cue ball into the red balls one at a time, so that the red balls fall into the pockets. Because of conservation of momentum, if the cue ball hits the red ball head-on, it will come to a stop and the red ball will start moving in the same direction. For example, in the table below, the player can make three straight shots to sink the red balls.

1. If the white cue ball stops every time it hits a red ball, can you can find a way to sink all the red balls, one at a time, for the table below?

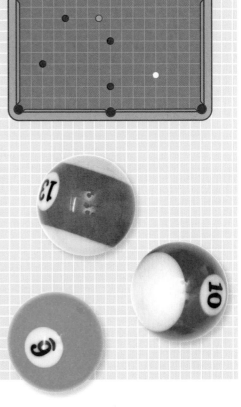

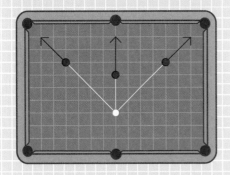

BOUNCING OFF CUSHIONS

When a ball hits a cushion, it bounces off at the same angle it hit the side. Pool players use this fact to sink balls they wouldn't normally be able to sink. The table below shows how a player might bounce off a cushion to sink a ball.

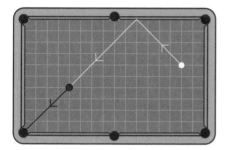

2. Can you find a way to sink all the red balls by bouncing off the cushions? The white ball should not hit any yellow balls.

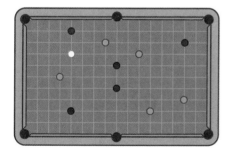

HITTING AT AN ANGLE

If the cue ball hits another ball at an angle, the two balls will move away from each other in a different way. Let's say you want to sink a red ball by hitting it at an angle. Before the collision, the cue ball will have momentum. Momentum is conserved, so you should expect the balls to have momentum after the collision. Suppose you hit the cue ball due east, and it

has a momentum of 5 units. After the collision, it moves with a momentum of 2 units due east, and 3 units due north.

3. What is the momentum of the red ball after the collision? There were 5 units of momentum due east before the collision, so there must be 5 units due east after the collision. The cue ball has 2 units of momentum, so the red ball must have 3 units of momentum due east:

Momentum before = Momentum after

5 units (cue ball before) + 0 units (red ball before) = 2 units (cue ball after) + 3 units (red ball after)

In the north-south direction, you know that the cue ball starts off with 0 units and ends up with 3 units due north. That means that the red ball must have 3 units due south, so that the total momentum in the north-south direction is zero:

0 units (cue ball before) + 0 units (red ball before) = 3 units (cue ball after) - 3 units (red ball after)

4. Using the same logic, can you work out what momentum the red ball has after the following collisions?

- The cue ball starts off with a momentum of 6 units due east. After the collision it has a momentum of 5 units due east, and of 2 units due south.
- The cue ball starts off with a momentum of 3 units due south. After the collision it has a momentum of 1 unit due east, and of 1 unit due south.
- The cue ball starts off with a momentum of 2 units due west and of 3 units due south. After the collision it has a momentum of 1 unit due west, and of 0 units due south.

DISCOVER: BASEBALLS AND CANNONBALLS

Have you ever looked at the shape made by water in a fountain? When water comes out of a spout that's at an angle, it has the shape of a smooth curve. That curve is called a parabola, and it's the shape objects follow when they are falling due to gravity.

BASEBALLS

When you hit a baseball, you want it to go as far as possible. What's the best way to hit it so that it travels as far as possible without touching the ground? As soon as you hit the baseball, the only force acting on it is gravity, which will make it accelerate toward the ground. You could hit it as hard as you can horizontally to make it travel a long way before it hits the ground, but is that really the best thing to do? If you hit it so it goes up slightly, it will take longer to hit the ground, so it could travel farther. If you take things to extremes and hit the baseball straight up vertically, it will stay in the air for the maximum possible time. (But it might then hit you on the head, and the point is to make it travel as far as possible.)

There must be some sweet spot in the middle. Making the baseball move horizontally keeps it in the air for the shortest time possible. Making it move vertically keeps it in the air for the longest time possible, but this also means it hits the ground at the same spot, so you may as well just drop the baseball. It turns out that the best angle to hit a baseball is at 45 degrees from the ground; 45 degrees is the perfect balance between giving the baseball as much horizontal speed as you can, and giving it more time in the air by giving it more vertical speed.

Baseball players spend years perfecting their swing, getting it just right.

CANNONBALLS

The reason why 45 degrees is the best angle to hit a baseball is because the baseball follows a parabola. This is the same for any object falling under gravity. Today, knowing how gravity works may give you an advantage on the baseball field, but for centuries this kind of knowledge could win or lose wars. Throwing objects up and letting them fall is known as projectile motion, and whatever is being thrown—whether it's a baseball or a cannonball—is called the projectile.

In the Middle Ages, cannons were used in battles to fire cannonballs at enemy castles. The goal was to damage the castle so that soldiers could get inside. This meant that physicists were very important to an army, and a lot of time and effort was spent trying to understand projectile motion.

As you can see in this diagram, the distance a cannonball will travel is controlled by the angle it makes with the ground. Armies that know how to hit their targets are better at winning battles than those that don't.

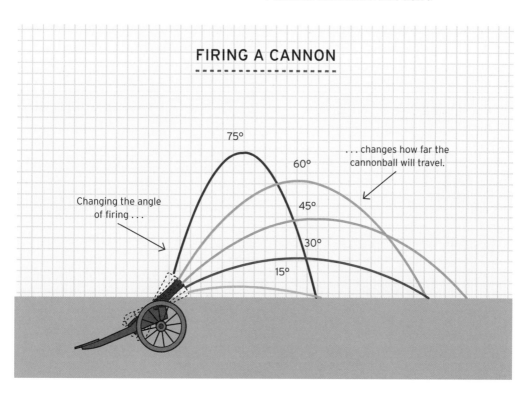

FIRING A CANNON

75°

60°

... changes how far the cannonball will travel.

45°

Changing the angle of firing ...

30°

15°

EXPERIMENT: POPCORN PROJECTILES

When an object falls under gravity, it speeds up. If you want to know how to hit a target with a slingshot, you need to know how it falls under gravity. In this experiment you'll find how to predict the movement of projectiles so that you can hit your target accurately.

YOU WILL NEED:

- Protractor
- Ruler
- Rubber band or small slingshot
- Unpopped and popped popcorn kernels
- Sticky tape
- Several cups for targets
- Nonpermanent way to mark the floor, such as a sticky note
- Large, heavy, solid object (a table leg, for example)
- Friend to help with marking where the projectiles land, and to play against

SETTING UP THE LAUNCHER

In this experiment, you'll be trying to hit some targets. This is a real-world problem that faced physicists and mathematicians when the cannon was invented, and firing projectiles became both vital to countries' defenses and very expensive. Remember, gravity can only affect vertical motion of an object, so its horizontal motion will stay the same.

WHAT TO DO:

To set up the experiment, you will need a launching station and targets.

1. To make the launching station, attach the protractor to the side of the large object with tape, so that it stays in place. The protractor should have the straight edge at the top and the semicircle at the bottom.

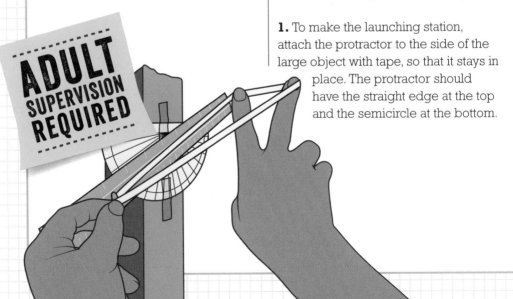

ADULT SUPERVISION REQUIRED

2. To fire a popcorn projectile, use the rubber band. One end of the rubber band should be at the center of the protractor, and then you can use the protractor to measure the angle, and the ruler to measure the extension of the rubber band (how far the rubber band is pulled).

3. Try firing your popcorn projectiles (using unpopped kernels) at different angles, including 0 degrees, 30 degrees, 45 degrees, 60 degrees, and 90 degrees from horizontal. Experiment with different extensions as well. The amount you extend the rubber band or slingshot will depend on how tight the rubber is. For each attempt, get your friend to mark where the projectile hit the floor and what the angle and extension were.

HITTING THE TARGETS

Now that you have this information, you can try to hit the targets. Lay out some cups as targets and, based on the information you have about how the projectiles move, see how many targets you can hit. See if you can have a competition with your friend to see how many targets each of you can hit with 10 attempts. The person who best understands the physics behind the motion will probably get the better score.

With some practice, you can become very good at this game, so to make things interesting, try using some popped pieces of popcorn. Try firing unpopped kernels and pieces of popped popcorn with the same angle and extension. Find the differences between where they land after multiple attempts. What conclusions can you draw? Does the popped projectile hit the ground closer or farther away from the launcher? How consistent is the distance traveled between popped projectiles with the same angle and extension? How does this compare to the unpopped kernels?

WHAT HAPPENS?

Firing popcorn at different angles changes how far the popcorn travels. Stretching the rubber band farther increases the speed of the popcorn, so that also affects how far the popcorn travels. Using unpopped kernels is easier because the shapes of different kernels are similar, and they have the same air resistance. Pieces of popped popcorn have very different shapes, so their air resistance is different. This slows them down by various amounts.

LEARN ABOUT:
ANGULAR MOMENTUM

If you spin a spinning top, it will continue to spin for a long time. Why does it do this instead of just stopping? It's because of something called angular momentum, which is conserved just like momentum is. Objects continue to spin unless and until something stops them.

WHAT IS ANGULAR MOMENTUM?

When objects move in a straight line they have momentum. In the same way, when they rotate they have angular momentum. Just like normal momentum, angular momentum is conserved, and that means it takes effort to increase or decrease how quickly something is rotating. If you lift a bicycle and spin its back wheel with the pedals, it will continue to spin after you stop pedaling. This is because the wheel is rotating, which means it has angular momentum. Angular momentum is conserved so the wheel continues to spin. The line that the wheel spins around is called the axis. There are three ways to change an object's angular momentum:

- Objects with more mass have more angular momentum.
- Objects that are rotating faster have more angular momentum.
- Objects that are more spread out from the axis of rotation have more angular momentum.

Imagine there is a spinning ice-skater. He decides he wants to spin faster, and he knows about angular momentum and how it is conserved. He decides he can either move his arms out (so he is more spread out) or move his arms in close to his body (so that he is less spread out). Which should he do to make himself rotate faster? A while later he is spinning again, and he notices his daughter. He picks her up, and they spin around together. Would you expect him to spin faster or slower after he has picked up his daughter?

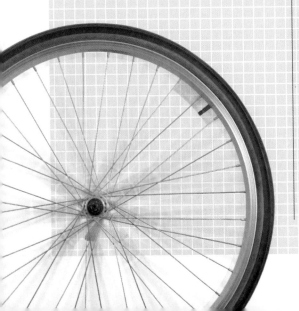

CHANGING EARTH'S ROTATION

Let's imagine there is a crazy villain who decides to change the way Earth rotates around its axis. He wants to make the day 30 hours long, so that people will work longer hours every day. To do this, does he need to speed up or slow down how much Earth spins around its axis? His assistant suggests four ways to slow down Earth's rotation, but only two of them will work. Out of the four suggestions, which two should the villain choose?

- Flatten Earth so that it looks like a very wide pancake instead of a sphere.
- Stretch Earth so that it looks like a broomstick, lined up with its axis of rotation.
- Hollow out the center of Earth so that it has less mass.
- Cover the surface of Earth with lead, so that it is very heavy.

The villain wants to make people work for 10 hours a days instead of 8 hours a day, and he claims this means people will work more in a year. What he doesn't know is that angular momentum also keeps Earth in orbit around the sun. If the year is 365 days long when the day is 24 hours long, how many hours are there in a year? If people work 8 hours a day for 5 out of 7 days, how many hours does the average person work per year? After the villain has made the day 30 hours long, how many new days fit into a year? If people work 10 hours per new day for 5 out of 7 new days, how many hours do they work per year? Has the villain actually made people work more per year?

DISCOVER: FRICTION AND RESISTANCE

It takes effort to keep moving around, to keep pedaling while on a bicycle, or to keep swimming to move in water. This is because there are forces that resist changes to motion. Friction is a contact force between solid objects, like bicycle tires and the road, and it always opposes movement.

HOW FRICTION WORKS

Friction is the force between two solid objects that move against each other. It's what makes you slow down when you ride your bicycle and stop pedaling. Friction can generate a lot of heat, so your tires will be warmer after you're done riding your bicycle than when you started. Some surfaces have more friction than others, so they are better at slowing you down. Ice has very low friction, which is why icy roads are more slippery and dangerous than ice-free roads.

Friction is also related to traction, which is what allows you to push against a surface. Bicycle tires are made of rubber, which has a lot of traction, so it's easy to push against the road in order to move—but that also means there is more friction against the road. Trains have metal wheels that run on rails. Both of these are made of metal, which has low friction and low traction. That means it takes a long time for a train to get up to full speed, but it experiences very little friction when it is at full speed. This also means that it is very hard to slow down a train quickly.

The friction an object feels depends on the surfaces that are touching and the mass of the object. Heavy objects will feel more friction than light objects. This is why it is harder to move a heavy box across a carpeted floor than a light box of the same size across the same floor.

It takes a long time to stop a fast train.

Air resistance can be a matter of life and death.

AIR RESISTANCE

Air resistance is the force of air resisting motion. It takes effort to move through air, although you don't normally notice it. This makes jogging down the street harder work than jogging on a treadmill, even though the speed and distance might be the same. The difference is that running through air takes more effort than running in place. Air resistance is usually bigger for faster objects and for objects that are larger. Parachutes use air resistance to slow down parachutists as they fall.

Moving through water is even harder than moving through air. This is because water is more dense than air, so there is more substance that has to be moved out of the way. This is one reason why the world record for running is about 12.4 m/s (27.8 mph), whereas the world record for swimming is about 2.2 m/s (4.9 mph).

Imagine a parachutist jumps out of an airplane. As they fall under gravity, their speed will keep increasing (see pages 84–87). However, this means that the air resistance will also increase because faster objects feel more air resistance. Eventually, the force of gravity will balance the air resistance, and the parachutist's speed will stay the same. This is called terminal velocity, and the terminal velocity depends on the object's mass and shape, as well as what substance it is traveling through. By opening a parachute, the parachutist changes their shape, as well as their terminal velocity, allowing them to land safely.

EXPERIMENT: FRICTION IN ACTION

What affects how quickly something slows down when there is friction and resistance? Is it mass, shape, speed, or the substance? It turns out the answer is a complex combination of all of these; some people spend their entire careers studying friction and air resistance.

YOU WILL NEED:

- 2 long, smooth planks of wood
- Stack of books, about 30 cm (1 ft.) high
- Several toy cars of different masses
- Several surfaces, including smooth wood, carpet, paper, and sandpaper
- Tape measure
- 2 glass pitchers of equal size
- Water
- Cooking oil
- 9 marbles
- 3 ball bearings

AIR RESISTANCE
WHAT TO DO:

1. Stack the books and place a plank of wood with one end on top of the books and the other on the floor, making a ramp.

2. Place the other plank at the bottom of the ramp for a racecourse. Place a toy car at the top of the ramp and release it. Measure how far it travels along the racecourse.

3. Replace the racecourse (bottom plank) with other surfaces and repeat the experiment.

On which surface does the car travel the most and least distance? Is this what you expected? What happens if you try with a heavier or lighter car?

FLUID RESISTANCE
WHAT TO DO:

1. Set up two pitchers next to each other on a table. Fill one pitcher with water and the other pitcher with oil. The liquids should reach the same height in both pitchers.

2. Take two marbles and—with one marble above the pitcher of water and the other outside it—drop them from the same height. In this way, you can compare the effect of air resistance to the effect of water resistance. Which marble reaches the bottom first? Is there more resistance from the air or water?

3. Repeat step 2, but this time use the pitcher of oil to compare the resistance of oil to air.

4. Compare the resistance of water to oil by dropping marbles into both pitchers at the same time. Which has the most resistance? Try other liquids, for example honey or cornstarch.

5. Now try dropping ball bearings and marbles into the same substance at the same time—for air, water, and oil. Which one reaches the bottom of the pitcher first?

WHAT HAPPENS?

With the toy cars, the main force acting on the cars is friction between the wheels and the racecourse. You will find that the rougher surfaces slow down the cars more quickly because they have more friction. Sandpaper and carpet have a lot of friction, whereas smooth wood and paper have less friction.

For the marbles in the air, water, and oil, the resistance depended on the speed and shape of the marbles, as well as the substance. Air is easier to move out of the way, so it has low resistance, and the marble moving through air should have hit the table before the other marbles hit the bottom of the pitcher. Water and oil are both liquids, so they have more resistance. Oil tends to be "thicker" than water, so it should slow down the marble more than water, but there can be quite a lot of variation depending on the type of oil.

Comparing marbles to ball bearings allows you to find out how shape and weight affect the resistance. The ball bearings are heavier than the marbles, so they have more weight, and their shape is usually smaller. This means that the force of gravity is larger and the force of resistance is smaller for ball bearings, and they'll reach the bottom of the pitcher faster than the marbles.

LEARN ABOUT: FRICTION AND TRUCKS

If you run along a boardwalk next to a beach, you can probably run quite fast. Running on sand is harder to do, and running in the ocean is even more difficult. This is because when you run on solid surfaces, you feel friction, and when you run in air or water, you feel resistance to movement.

Sometimes big trucks lose control when traveling down steep hills. This can happen because gravity pulls the truck down, but the brakes on the truck, and friction between the tires and road, are not enough to balance the effect of gravity. To avoid serious accidents, some roads have special escape lanes so that trucks can come to a stop safely. These escape lanes work because their surface is made from high-friction materials.

TRUCKS AND ESCAPE LANES

Imagine there is a truck that has just come down a steep hill in icy conditions and is now traveling on a horizontal stretch of road. The truck is traveling at 30 m/s, and the driver wants to slow the truck down quickly. She applies the brakes, which can slow down the truck by 1 m/s each second. She now has a choice of three different routes:

- She can choose to continue to travel on the road, which slows down the truck by an additional 1 m/s each second, but there are only 200 m before she has to stop at the next stop sign.
- She can choose to use the escape lane, which slows down the truck by an additional 3 m/s each second, but it is only 50 m long.
- She can drive onto a gravel surface, which slows down the truck by an additional 2 m/s each second, but it is only 175 m long.

If she chooses to travel on the road, she needs to take her speed down from 30 m/s to 0 m/s. To work out how far she would travel, we need to know her average speed. To get the average speed, we use the equation: Average speed = (Speed at start + Speed at end) ÷ 2 = (30 + 0) ÷ 2 = 15 m/s. The distance the truck will travel in this time cannot be more than 200 m if it is to stop before the end of the road.

Using Speed = Distance ÷ Time, this means the amount of time she has available is:

200 m ÷ 15 m/s = 13 seconds. Is that enough time?

Every second, her brakes can slow down the truck by 1 m/s, and the friction gives her another 1 m/s. This gives a total of 2 m/s every second.

Remembering the equation for acceleration (Change in speed = Acceleration × Time), this means she can change her speed by 2 m/s each second × 13 seconds = 26 m/s.

This is not enough to stop her truck. If she reduces her speed (which was 30 m/s) by 26 m/s, then she would be traveling at 4 m/s by the time she reached the stop sign. So she should not choose to stay on the road.

Using the same reasoning, which of the other two options should she choose?

CHAPTER 4
NUCLEAR PHYSICS AND SPACE

DISCOVER...

LEARN...

EXPERIMENT...

DISCOVER: ATOMS AND THE NUCLEUS

Everything you see and touch is made up of atoms. Atoms are tiny objects, too small to be seen. For a long time, scientists thought that atoms were the smallest things that existed, but it turns out there are even smaller things inside atoms.

LOOKING INSIDE AN ATOM

A particle is a small object that looks like a tiny ball. Most of the time, atoms look like particles, but if you could look inside an atom, you would see lots of smaller things inside. Most of an atom is empty space. At the center of the atom is the nucleus. The nucleus is made up of two other kinds of particles called protons and neutrons, which take up less than 0.000000000001% of the space inside an atom, but make up 99.9% of the mass of the atom. The rest of the atom is made up of particles called electrons. These electrons whizz around in a big cloud, taking up the other 99.999999999999% of the space.

To get an idea of the scale, imagine an empty soccer field. If the whole soccer field is the size of an atom, then a popcorn kernel at its center is the size of a nucleus.

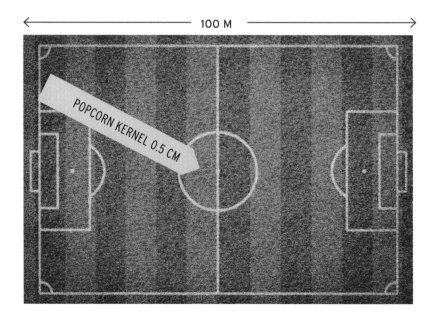

100 M

POPCORN KERNEL 0.5 CM

The protons have a positive electric charge, the neutrons have no electric charge, and the electrons have a negative electric charge. These charges are what keep the atom together. The protons and electrons have opposite electric charges, so they attract each other. The nucleus has only positive electric charges. When objects have the same electric charge, they repel (push away) from each other, so how does the nucleus stay together?

GOING FISSION

There are two other forces inside the nucleus that help to keep it together. These two forces are called the strong nuclear force and the weak nuclear force. These forces are usually stronger than the electric force, so the nucleus stays together. Sometimes the forces are nearly balanced, and if that happens, the nucleus can split. When a nucleus splits, it is called fission, and it can release a lot of energy. This energy is what powers nuclear power plants.

A nuclear power plant uses atoms that are ready for fission, and the best atom for this is a uranium atom. A uranium atom has 92 protons and 143 neutrons. If you fire a neutron at a uranium atom, its nucleus will split and release three neutrons. These neutrons can hit some more uranium atoms, and they can split too. This is called a chain reaction, and it can

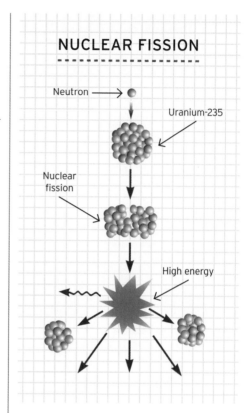

NUCLEAR FISSION

Neutron

Uranium-235

Nuclear fission

High energy

keep going as long as there are still uranium atoms for fission.

Fission releases a lot of energy as heat, and this heat can be used to make steam. The steam can turn a turbine, and that can be used to generate electricity. The amount of energy released is so great that 1 kg (2 lb.) of uranium can release the same amount of energy as 2,500,000 kg (2,755 tons) of coal.

EXPERIMENT: NUCLEAR POPCORN

The nuclear forces can have lots of different effects. Sometimes the forces inside a nucleus are not balanced, and when this happens, the nucleus is unstable. An unstable nucleus can emit (throw out) a particle, and then it can become a stable nucleus. Popcorn behaves like this when it's cooked.

MODELING NUCLEAR DECAYS

An unstable nucleus can emit a particle to become stable, and this is called nuclear decay. Even though you don't know when a nucleus will emit a particle, if you have enough of them, you can work out how many will emit a particle. The amount of time it takes half of them to emit a particle is called the half-life. Different kinds of nuclei have different half-lives. A very unstable kind of nucleus can have a half-life of a fraction of a second, and a very stable kind of nucleus can have a half-life of thousands of years.

You can use microwave popcorn to demonstrate what nuclear decays are like.

YOU WILL NEED:

• 4 bags of microwave popcorn
• Microwave oven
• Stopwatch
• Kitchen scale
• 1 red pen and 1 blue pen

WHAT TO DO:

1. Take a bag of microwave popcorn and place it in the microwave. Press the start button.

2. Once you hear the first pop, start the stopwatch. Stop the microwave after 20 seconds.

3. Carefully open the bag and let it cool down for a few seconds. Separate out the popped pieces of popcorn from the unpopped kernels.

4. Count how many popped pieces of popcorn, and how many unpopped kernels, there are. Write down the fraction of popped pieces of popcorn and the fraction of unpopped kernels. For example, if you have 30 pieces of popped popcorn, and 20 unpopped kernels, then the total number is 50. The fraction of popped popcorn is 30 popped kernels ÷ 50 = 60%; the fraction of unpopped kernels is 20 unpopped kernels ÷ 50 = 40%.

5. Repeat this for 50, 80, and 110 seconds after the first pop, using a new bag of popcorn each time.

LOOKING AT THE RESULTS

When you have the fractions, copy the graph below and draw a cross for each fraction. Use a red pen for the fraction of popped pieces, and a blue pen for the fraction of unpopped kernels.

Draw a smooth curve through the red points, and another smooth curve through the blue points. The curves should cross when the fractions are at about 50% of the pieces of total popcorn. Draw a line down from where the curves cross, and see at what time this happens. This is the half-life of the popcorn. After this amount of time, about half the kernels will have popped. There is an example graph in the answers section to show you roughly what the graph should look like.

WHAT HAPPENS?

When the bag of popcorn gets hot enough, the popcorn starts to pop. This is because the forces inside the kernels get stronger than the forces keeping the kernels together. You can't tell which kernel will pop next, but with enough kernels, it's possible to predict how many will pop in the next minute. This is just like how nuclear decays work. Both popcorn and nuclear decays have their own half-life.

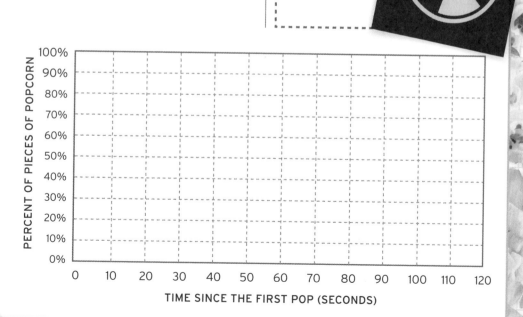

LEARN ABOUT: INSIDE AN ATOM

Atoms are so small that they cannot be seen. If you could see inside atoms, things around you would look very different than how you normally see them. Even if you hold a solid object like a rock, on the scale of an atom, it looks like it's mostly empty space.

EXPONENTS

Sometimes you'll need to use very large and very small numbers. For example, a billion is 1,000,000,000—it can be tedious to count all those zeroes, so it's easier to write that as a power of 10. You can write 1 billion = 1,000,000,000 = 10^9, where the 9 tells you that there are nine zeroes. In other words, to get 1 billion you need to need to multiply 1 by 10 nine times. In the same way, a billionth is 0.0000000001, which can be written as 1 billionth = 0.0000000001 = 10^{-9}. Again, there are nine zeroes, but the negative symbol means that you are looking at a very small number instead of a very large number. The power of 10 is called the exponent.

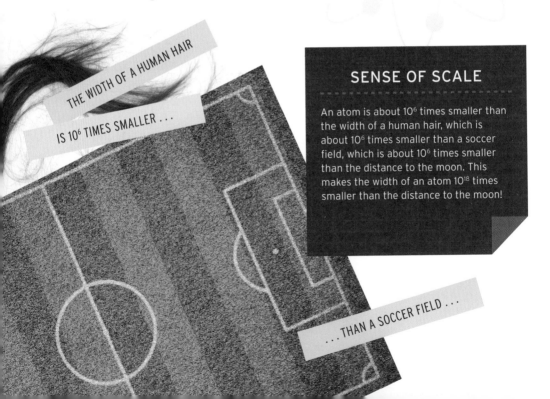

THE WIDTH OF A HUMAN HAIR

IS 10^6 TIMES SMALLER ...

SENSE OF SCALE

An atom is about 10^6 times smaller than the width of a human hair, which is about 10^6 times smaller than a soccer field, which is about 10^6 times smaller than the distance to the moon. This makes the width of an atom 10^{18} times smaller than the distance to the moon!

... THAN A SOCCER FIELD ...

. . . WHICH IS 10^6 TIMES SMALLER THAN THE DISTANCE TO THE MOON.

VERY BIG AND VERY SMALL NUMBERS

When numbers with the power of 10 form are multiplied, the exponents are added up. Try multiplying 1 million by 1 thousand:

- First, write 1 million in powers of 10:
  1 million = 1,000,000 = 10^6
- Then write 1,000 in powers of 10:
  1 thousand = 1,000 = 10^3
- Multiply them by adding the exponents:
  $10^6 \times 10^3 = 10^{6+3} = 10^9$
- To check that you have the right answer, you can turn the exponents back into zeroes:
  10^9 = 1,000,000,000 = 1 billion

When dividing numbers in the power of ten form, you subtract the exponents. For example, dividing 1 million by 1 thousand:
$10^6 \div 10^3 = 10^{6-3} = 10^3 = 1,000.$

When multiplying or dividing numbers in the power of 10 form, you multiply or divide the rest of the numbers in the normal way. Let's divide 8 million by 2 thousand:
$(8 \times 10^6) \div (2 \times 10^3) = (8 \div 2) \times 10^{6-3} = 4 \times 10^3$
= 4,000.

DIFFERENT DENSITIES

A hydrogen atom has one proton and one electron. The proton makes up the nucleus, and it takes up 10^{-15} of the space in the atom. If the mass of the proton is 1.7×10^{-27} kg (1.9×10^{-30} tons), and the space inside an atom is 6×10^{-31} m³ (2×10^{-29} cu. ft.), what is the density of a proton? The equation for density is:

Density = Mass ÷ Volume

Don't forget that the proton only takes up a tiny fraction of the total volume of the atom.

Electrons are tiny, and they move around in a cloud. This cloud takes up the rest of the space in the atom. The electron has a mass of 9.1×10^{-31} kg (10^{-35} tons). What is the density of the electron in this cloud? *Hint: you can assume that the electron cloud takes up all the space of the atom.*

How much more dense is the nucleus (proton) compared to the electron cloud?

DISCOVER: MEDICAL PHYSICS

What would it be like to look inside a human body? Using the tools of modern physics, it is possible to do just that. You have probably seen X-ray images in real life or on TV. Scientists and doctors look inside objects, including the human body, using nuclear physics, electromagnetism, and even the physics of sound.

X-RAY IMAGES

X-rays are actually a kind of electromagnetic radiation. X-rays are just like visible light, except they have a shorter wavelength. The wavelength of an X-ray is about the same as the size of an atom, and that means that X-rays bounce off atoms. It's possible to fire some X-rays at an object, see how they bounce off, and use this information to work out what the object looks like.

Bones contain a lot of calcium, and calcium is great at bouncing X-rays. So X-rays are very good at seeing inside a human body and seeing where the bones are. If someone breaks a bone, their doctor will usually take an X-ray image to see what the damage looks like and how well it is healing. Metal is also great at bouncing X-rays, so they can be used to look at artificial hips and other medical equipment. Being exposed to too many X-rays can be bad for your health, though, so the number of times someone has an X-ray has to be controlled.

PET IMAGES

PET (positron-emission tomography) scans are another way for doctors to look inside the human body. A positron is an electron with a positive electric charge, and tomography means looking inside an object. When a positron meets an electron, they join

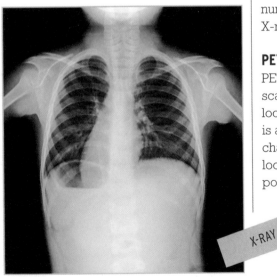

X-RAY

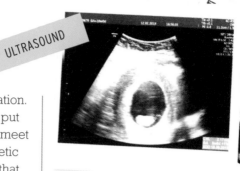

ULTRASOUND

up to make electromagnetic radiation. That means that if positrons were put inside a human body, they would meet electrons and make electromagnetic radiation. If you could see where that radiation came from, you would be able to see where the positrons were. This is how PET works.

Oxygen-15 can release a positron, and oxygen can be found in water. So, some water with oxygen-15 can be injected into a person, and then PET can be used to make a map of where that water goes. PET can be used to find diseases such as cancer and to measure how blood moves in the brain and heart.

MAGNETIC RESONANCE IMAGES

Magnetic resonance imaging (MRI) uses magnetic fields to make a map of the body. These images are good at making maps of where the water and fat in a body is. This allows doctors to look at the softer parts of bodies, such as organs in patients, which can help find lots of kinds of injuries and diseases.

ULTRASOUND IMAGES

Ultrasound images use very high-frequency sound waves to make a map of the body. The sound waves echo off parts of the body, and these can be used to make an image. Making ultrasound images is safe, so ultrasound is often used for looking at very sensitive parts of the body or making lots of images. For example, ultrasound is used to see how babies develop before they are born.

MRI

EXPERIMENT: X-RAY SCANNERS AND BAGS

X-ray images are used to scan bags for banned objects. Using X-rays is a fast way to look inside a bag and find hard objects. They can be used to find jewels and other precious items, but it can be difficult to be fast and reliable with lots of bags.

YOU WILL NEED:

- Envelopes
- Several sheets of paper
- Scissors
- Powerful flashlight
- Sheets of thin cardboard (optional)

WHAT TO DO: SETTING UP THE BAG CHECK

Imagine that the local bank has been robbed, and you are a TSA agent asked to check several hundred bags at the airport for missing diamonds. Would you be able to do it quickly enough?

1. Draw twenty shapes on paper of items that someone might take on vacation. These could be clothes, snorkels, flip-flops, and so on.

2. Draw the shape of three diamonds that have been stolen from the bank.

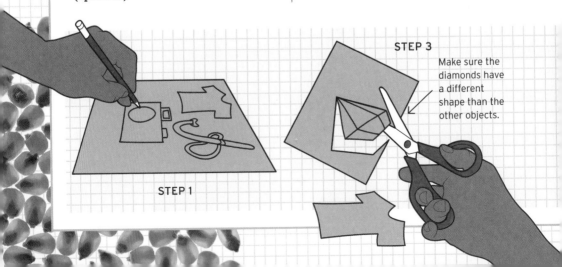

STEP 3

Make sure the diamonds have a different shape than the other objects.

STEP 1

3. Cut out all these shapes and place them into two piles of "normal" and "stolen" items. Take five envelopes to use as bags and put the items in them in any way you like. Shuffle the "bags" and lay them on the table in front of you.

4. You will try to find the stolen items, and the only tool you have to help you is the flashlight. The flashlight is your X-ray source. Take a bag, shine the X-ray source through one side of the bag, and look at the other side. Can you find the stolen items by doing this?

BEAT THE SYSTEM

To make things more interesting, play a game with two or more friends.

1. Each player prepares 20 normal items and three stolen diamonds.

2. Each player should take five bags (envelopes) and label them *A, B, C, D, E*. These bags should be filled with items in any way the players like.

The players should write down (in secret) which bags contain the stolen items. Next, they shuffle the bags and pass them to the player on their left.

3. Players should take turns using the flashlight as an X-ray source to scan the bags. They should write down where they think the stolen items are. When everyone has decided, they should compare this to where the stolen items really are. The player who locates the most stolen items wins.

If you find this too easy, play the game again, but time how long each player takes. In the case of a tie, the player who finds the stolen items in the shortest time wins. To make things even more challenging, use thin cardboard for some of the non-stolen items.

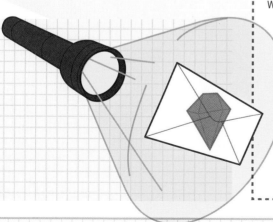

WHAT HAPPENS?

When you shine the flashlight through the envelopes, the sheets of paper absorb some of the light. This leaves a shadow, and you can see the shapes of the items inside the bag. When the shapes overlap, the shadows become darker. With X-ray images, objects that are denser are better at bouncing X-rays, so they look clearer, and this is like having more layers of paper.

LEARN ABOUT: RADIATION AND ORANGES

Radiation is all around us, including natural radiation from the ground and from outer space. Even human beings are slightly radioactive! The most common kinds of radiation are called alpha, beta, and gamma radiation.

RADIATION TYPES

There are many different types of radiation, and each type is made of different kinds of particles.

Alpha radiation is the same as the nucleus of a helium atom. It has two protons and two neutrons. Alpha radiation is usually quite easy to absorb, and a single sheet of paper is normally enough to stop it.

Beta radiation is actually an electron. It takes a bit more to stop beta radiation, but a sheet of metal will normally be strong enough.

Gamma radiation is a form of electromagnetic radiation with a very short wavelength. A very short wavelength means a lot of energy. It takes a lot to stop gamma radiation, and usually very dense materials like lead are used to do this.

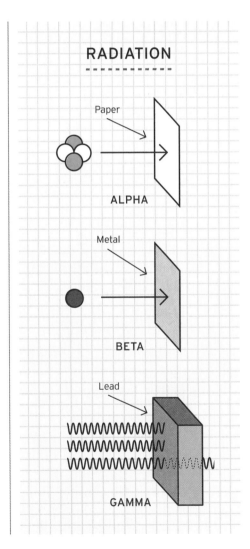

RADIATION

Paper

ALPHA

Metal

BETA

Lead

GAMMA

ORANGE EQUIVALENT DOSE

Oranges are naturally radioactive, and they give off beta and gamma radiation. The amount of radiation emitted by an orange is not dangerous, but it can be used to measure how much radiation there is around us. For example, taking a flight from London to New York would be the equivalent of eating 800 oranges in terms of the amount of extra radiation you would receive. The "safe" limit for radiation is the equivalent of 20,000 oranges per year, which is a bit more than two oranges per hour.

BLOCKING RADIATION

1. Imagine you are sitting next to an alpha radiation source, a beta radiation source, and a gamma radiation source. Each one has an orange equivalent dose of two oranges per hour. After one hour, what would be the orange equivalent dose of sitting next to these sources of radiation? What about after one day? What would be the orange equivalent dose after one year? Is that more or less than the recommended limit of 20,000 per year?

2. You decide to put some protection between yourself and the radiation. If you put a piece of paper between you and the radiation, would that be enough protection? What if you also put a sheet of metal between you and the radiation? Assume that 1 cm (about 0.5 in.) of lead stops half of all the gamma radiation. How much lead would you need to use to reduce the total radiation from all three sources to about a 2,000 orange equivalent dose per year?

FUN FACT: Gamma radiation is sometimes used to kill harmful bacteria on fruit and vegetables, making them safer to eat.

DISCOVER: ENOUGH POPCORN TO MAKE A STAR

How much popcorn would be needed to make a star? That might sound like a strange question, but if you bring enough atoms together, they will eventually form a star. A star is just a huge amount of mass that can make its own heat and light.

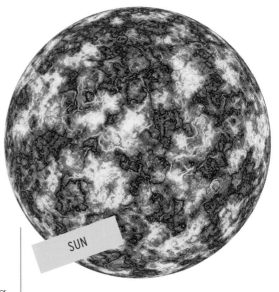

SUN

HOW THE SUN FORMED

The sun is huge and is roughly 300,000 times heavier than Earth. Most of the sun is made up of hydrogen atoms (in fact, most of the universe is made up of hydrogen atoms). The sun wasn't always shining, and in the beginning it was just a cloud of hydrogen gas. Even though each hydrogen atom is extremely light, it still has a gravitational force. When there are a large number of atoms, the gravitational force is strong enough to pull them together. This is what made the gas cloud form together into a big ball.

The weight of the gas had been pushing down on all the gas below it, which increased the pressure and temperature. At the center of the sun, the atoms were smashing into each other, and occasionally, the protons would hit each other. Normally, protons repel each other because they all have the same electric charge, but at the center of the sun, the force pushing them together was a lot stronger than the force repelling them.

FUSION

Every now and then two protons would hit each other and make a neutron. Then these neutrons would hit other protons. If these protons and neutrons hit each other just right, they would make the nucleus of a helium atom— that's two protons and two neutrons stuck together. Turning hydrogen into helium like this is called fusion, and fusion releases a huge amount of energy. This is what keeps the sun shining.

As long as a star has enough hydrogen for fusion, it will keep shining. If the star is heavy enough, when it runs out of hydrogen, other atoms will fuse, and this will make even heavier atoms. This will continue until, billions of years later, the star will explode in what is called a supernova, sending out heavy atoms across the universe. Nearly all the atoms on Earth were originally made in the centers of old stars, which exploded more than 4 billion years ago. Some atoms in your left hand could have come from a different star than the atoms in your right hand!

SUPERNOVA

MAKING A STAR FROM POPCORN

How many kernels of popcorn would be needed to make a star? It doesn't matter what is used to make the star— what matters is how much mass there is. This is because if there is enough mass, fusion will start at the center, and a star will form. The smallest mass required to make a star is about 1.6×10^{29} kg. If a single kernel of popcorn is 0.16 g, then it would take 10^{33} kernels to make a star. That's a million, billion, billion, billion kernels. That would be enough to supply everyone on Earth with enough popcorn to eat a bag a day for a trillion years.

EXPERIMENT: MAKING A SOLAR SYSTEM

It can be hard to appreciate just how small Earth is compared to the rest of the universe. There are many planets, moons, and other objects in our solar system, but it is mostly empty space. The distances are so vast that it can be very difficult to imagine them, but making a model will help you to visualize things more clearly.

YOU WILL NEED:

- Popped popcorn and unpopped kernels
- Sticky notes
- Pen
- Grain of salt
- Orange
- Tape measure (able to measure in centimeters)

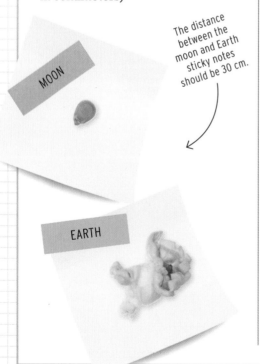

The distance between the moon and Earth sticky notes should be 30 cm.

MOON

EARTH

EARTH AND THE MOON

Earth and the moon interact with each other because of gravity. The force between them is so strong that it makes the moon egg shaped, and the same side always faces Earth. Earth is about 3.7 times larger than the moon, which is about the same ratio of a piece of popped popcorn to an unpopped kernel. Take two sticky notes and label one as Earth and one as Moon. Place a piece of popped popcorn (about 1 cm in diameter) on the Earth sticky note and an unpopped kernel (about 0.3 cm in diameter) on the moon sticky note.

The distance between Earth and the moon is about 30 times larger than the width of Earth. To get a sense of scale of Earth and the moon, move the moon sticky note to 30 cm away from the Earth sticky note. This will show you how much space there is between Earth and the moon and how small Earth is compared to this distance.

OUR SOLAR SYSTEM

Earth moves around the sun because of gravity (in the same way that the moon moves around Earth). The sun is about 100 times larger than Earth, and the distance between the sun and Earth is about 12,000 times larger than the size of Earth. To put Earth and the sun to scale, place a single grain of salt (about 0.5 mm in diameter) on a sticky note for Earth, and an orange (about 5.5 cm in diameter) on a sticky note for the sun. From this you can see how tiny Earth is compared to the sun. Move the Earth sticky note until it's 6 m away from the sun sticky note.

The largest planet in the solar system is Jupiter. It is about 10 times larger than Earth, and it's 5 times farther from the sun than Earth is. To add Jupiter, place 4 popcorn kernels in a square shape (about 5.5 mm wide) on a sticky note and place it 39 m from the sun sticky note. Your home might not be big enough for this. You can look up the sizes and distances of the other six planets and then work out where you would need to place them and how big they would be compared to Earth.

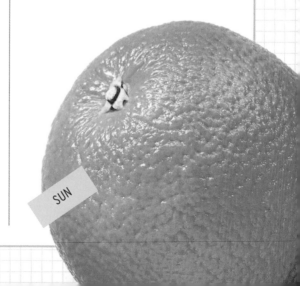

LEARN ABOUT: ESCAPE VELOCITIES

In 1969, two astronauts stepped onto the surface of the moon for the first time. They got to and from the moon on a rocket, and that rocket is the fastest vehicle that ever carried humans. Why did they have to travel so fast to get to the moon?

ESCAPE VELOCITIES

Earth is very heavy, and has a mass of about 6×10^{24} kg. This mass generates gravity, which is felt as weight. The weight of an object depends on how far it is from Earth. If you held an apple while standing on the surface of Earth, it would feel just like a normal apple. However, if you traveled in space as far away from Earth as the moon is, that same apple would feel 3,600 times lighter. Its weight would be 3,600 times smaller because the distance from the center of Earth would be much larger. If you put the apple as far away as the sun, its weight would be a billion times smaller.

As the apple gets farther away, the effect of gravity gets smaller. Let's say you threw the apple up into the air. Earth's gravity would slow it down, and it would eventually stop and then fall back to Earth. What if you threw the apple more quickly? It would reach even higher before it fell down. If gravity becomes very weak at very large distances, would it be possible to throw the apple quickly enough to

never come back down? In principle, yes. The speed at which you have to throw the apple (or any object) is called the escape velocity.

The escape velocity of Earth is huge—about 11,135 m/s (24,900 mph). The escape velocity for different planets depends on two things. If there are two planets of the same size but different masses, the planet with more mass has a larger escape velocity. If there are two planets of the same mass, but different sizes, the planet that is smaller in size has a larger escape velocity.

ESCAPING OTHER PLANETS

Imagine you are on a mission from Earth to a planet called Romulus. Romulus has the same radius as Earth, but four times the mass. Is the escape velocity of Romulus larger or smaller than the escape velocity of Earth?

Next, you travel to Remus, which has the same mass as Earth, but has nine times the radius of Earth. Is the escape velocity of Remus larger or smaller than the escape velocity of Earth?

Finally, you travel to Vulcan, which has nine times the mass of Earth and one-quarter the radius of Earth. Is the escape velocity of Vulcan larger or smaller than the escape velocity of Earth?

Here are the escape velocities for a few objects:

- A kernel of popcorn: 1.5×10^{-6} m/s (3.5×10^{-6} mph)
- An elephant: 9×10^{-4} m/s (2×10^{-3} mph)
- The moon: 2.4×10^{3} m/s (5.3×10^{3} mph)
- The sun: 2.8×10^{5} m/s (6.2×10^{5} mph)
- A neutron star, one of the heaviest objects in the universe: 1.6×10^{8} m/s (3.5×10^{8} mph)
- A black hole, the heaviest object in the universe: impossible. You cannot travel fast enough to escape a black hole.

DISCOVER: HOW ROCKETS AND SPACESHIPS WORK

Far above Earth is the International Space Station (ISS). People from all over the world live and work there, and rockets are used to deliver supplies of food and other essentials. It's often said that rocket science is difficult, but the ideas that keep the rockets going to and from the space station are actually simple.

MAKING A ROCKET LAUNCH

Rockets have to get to very high speeds to launch. They need to travel about 1,000 times faster than a car, and this takes a huge amount of energy. The way a rocket works is quite easy to understand. Newton's Third Law states that every action has an equal and opposite reaction. If you were on ice skates and you threw a heavy ball forward, you would move backward because of Newton's Third Law.

Rockets work in the same way, but they move vertically instead of horizontally. They burn fuel and eject it down very quickly. As this happens, Newton's Third Law means that the rocket moves up with an equal force. If enough fuel is ejected down quickly enough, the rocket will continue to move up. If the rocket reaches the escape velocity

of 11,135 m/s, it will be able to escape Earth's gravity.

Getting to the right speed in one step is not a very efficient way to get into space, and space missions actually use lots of different stages to launch a rocket. As the rocket burns fuel, it loses mass, so after a few minutes, a lot of the mass has been used up. Parts of the rocket are removed and fall back to Earth to a safe location. The rocket will continue to burn fuel, and with less mass, the rocket can reach the speed it needs more quickly.

ON THE ISS

The space station moves in a very special way. It moves around Earth so quickly that, even though it is always falling toward Earth, it never gets any closer to it. To understand how this works, imagine a cannon on the top of a large mountain. If you fire a cannonball, it will fall and eventually hit Earth. If the cannonball goes faster, it will travel farther before it hits Earth. You can control how far it will travel by controlling how fast it goes. If the cannonball is fast enough, it will go all the way around Earth and never land. That's how the International Space Station stays orbiting Earth at the same height. It is moving quickly enough sideways that the effect of this movement exactly cancels out the effect of Earth's gravity.

This makes astronauts feel like they are weightless. They can "swim" around the station, and spin around so that up and down become meaningless to them. If they decided to make their own popcorn and opened the bag, the popcorn would float around freely. It can take a while to get used to eating, drinking, and cleaning in space. The astronauts have to do special exercises to keep their bodies healthy because they cannot feel the effects of gravity.

EXPERIMENT: BUILD YOUR OWN SPACE SHUTTLE

For 30 years, the National Aeronautics and Space Administration (NASA) used space shuttles to fly missions into space. The world tuned in to watch these giant machines launch into space, with astronauts and scientists aboard. Here, you will learn about the various parts of a shuttle by making your own model, using only everyday kitchen items.

A space shuttle is made up of four main components: the vehicle that contains the astronauts and equipment, the two main rockets, and the fuel tank. The rockets work by burning fuel and expelling hot gas, which pushes the shuttle upward.

YOU WILL NEED:

- Empty 2-liter soda bottle
- 2 paper towel tubes
- 2 empty yogurt containers (clean them first)
- Empty 1-liter milk or juice carton
- A few sheets of cardboard
- 7.5 cm (about 3 in.) of sticky tape
- Paper
- Felt-tip pens
- Paper plate
- Paint (optional)
- Scissors

WHAT TO DO:

The model of the space shuttle will consist of the vehicle, the fuel tank, and the rockets.

1. To make the fuel tank, you'll need the large soda bottle. You can cover it with paper and paint it, if you want to make it look more realistic.

2. For the rockets, cover the paper towel tubes with paper.

3. Attach one yogurt container to the bottoms of each of the rockets to make the thrusters, which is where the hot gas comes out.

4. Cover the tops of the rockets with paper cones. You can make a paper cone by cutting out a semicircle and folding it around, using tape to hold its shape. The base of the cone can be taped to the top of the rocket.

5. For the vehicle, cover the milk or juice carton with paper. You can decorate the paper with the name of the shuttle and the US flag.

6. Cut out wings from the cardboard and tape them to the sides of the vehicle. Cut out a tail fin from the cardboard, and tape it to the back of the vehicle, as shown below.

7. Attach the vehicle to the top of the fuel tank using a 2.5-cm (about 1-in.) loop of sticky tape. Attach the rockets to the left and right sides of the fuel tank in the same way.

LAUNCHING THE SHUTTLE

Use an upturned paper plate as a launchpad. Place the completed shuttle on the launchpad and simulate the launch by lifting the space shuttle. The first parts to be detached are the rockets. Finally, when the shuttle is nearly in orbit, the fuel tank is detached. That leaves the shuttle in orbit, until it is ready to return to Earth.

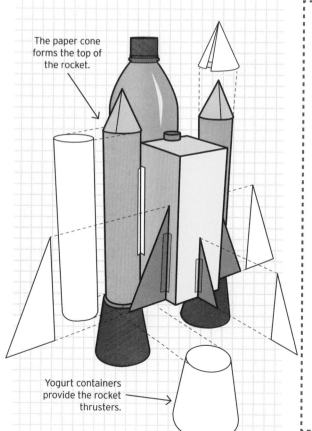

The paper cone forms the top of the rocket.

Yogurt containers provide the rocket thrusters.

WHAT HAPPENS?

The shuttles were extremely heavy, and NASA worked hard to reduce the weight of a shuttle during the launch, which is why the rockets and then fuel tank are detached. The shuttle is designed to meet as little air resistance as possible during the launch, whereas the vehicle is designed to return to Earth.

In order to reduce its speed when returning to Earth, the vehicle will fly backward and upside down for some time. As it gets closer, it flips the right way around. Just before it touches down, it extends some wheels and lands on a runway, just like an airplane. It is then returned to NASA by attaching it to the back of a large airplane.

LEARN ABOUT:
SPACE MISSIONS

Scientists have been sending spacecraft into space for more than 60 years. These space missions provide information about the universe, and some of them carry information about the human race, in case they are found by aliens. There are still a lot of interesting questions that can be answered with experiments in space.

SPACE EXPERIMENTS

There have been plenty of space-based experiments, with some landing on different planets, and others traveling around Earth. Each one tells scientists something different about the universe and helps them to understand our place in it a bit more. Successful space missions are hard to achieve, so each one is usually one small step up from the last one.

Can you match the names of the missions to what they studied? *Hints: it's easier to fly into space than to land on the moon, and it's easier to fly to the moon than to another planet. Landing a vehicle on another planet is harder than flying to it, and flying to a comet or an asteroid is harder still. The farthest human-made object from Earth has been traveling for almost 50 years.*

1. Sputnik (1957, USSR)

2. Pioneer 1 (1958, USA)

3. Vostok 1 (1961, USSR)

4. Mariner 2 (1962, USA)

5. Apollo 11 (1969, USA)

6. Voyager 1 (1977, USA)

7. Giotto (1985, EU)

8. Mars Pathfinder (1996, USA)

9. International Space Station (1998, International)

10. Hayabusa 2 (2014, Japan/Germany/France)

11. Europa Clipper (set to launch by 2025, USA)

SPACE MISSIONS

a) The satellite for this mission has traveled more than 13 billion miles, and it is the farthest humans have ever sent an object from Earth. Signals are still received from it, even though it has left our solar system.

b) After almost a decade of the Space Race, this mission successfully put the first humans on the moon. Neil Armstrong and Buzz Aldrin walked on the surface of the moon for more than two hours.

c) Halley's Comet is a huge rock that orbits the sun once every 75 years. This mission flew by Halley's Comet to make close-up observations.

d) This mission launched the first-ever satellite into space. It gave off a constant radio signal, which sounded like a beeping sound when it was detected on Earth.

e) This mission launched a station where astronauts could live in space and take part in a variety of experiments. It orbits Earth every 90 minutes. It would not be possible without a lot of international collaboration.

f) This is a planned mission to Europa, one of Jupiter's moons. Europa is thought to contain oceans of water, and there may be simple life on this moon.

g) This mission was the first rover on another planet. A rover is a special vehicle for traveling on other planets and moons.

h) This was the first mission where a human was sent up into space and made an orbit around Earth. The astronaut, Yuri Gagarin, was made a national hero for being the first man in space.

i) This mission had the first spacecraft to visit another planet, Venus. Venus is similar to Earth in size, but it has a runaway global warming problem, making it the hottest planet in our solar system. Its atmosphere contains acid, and it is a very dangerous place, even for robots.

j) This mission had the first spacecraft launched by the United States. It never reached the moon, but it did study the magnetic fields around Earth.

k) This was the first mission to land on an asteroid (a large rock orbiting the sun) and leave with a sample. It is due to return to Earth in 2020.

DISCOVER: SPECIAL RELATIVITY

Have you ever sat on a train, looked out of the window, and seen another train moving? It can be hard to tell if your train is moving and the other train is standing still, or if the other train is moving and your train is standing still.

FISH AND FIRE TRUCKS

In the 1500s, Italian scientist Galileo was interested in how things moved. People used to think that an object would only move if you kept pushing it, but Galileo realized that if there is no friction, moving objects continue to move. This is like when an ice hockey puck moves on ice; it will keep moving because the ice is very smooth. Galileo took this a step further and imagined a ship sailing on calm water. He thought about a goldfish swimming in a bowl on the ship. If the goldfish could not see outside, would it know if the ship was sailing on calm water or sitting stationary at a dock? If you are moving very smoothly, you cannot tell if you are moving at all without looking outside. This is an example of what is called relativity, and for a long time, it was thought to be a very simple idea, and quite boring.

However, this all changed at the start of the twentieth century, when physicists discovered that light always travels at the same speed in empty space: 299,792,458 m/s (670,616,629 mph). German physicist Albert Einstein was fascinated by this fact, and he thought about it in terms of riding on a fire truck. Imagine you are on a fire truck and you can spray water out of a hose at 5 m/s (11.2 mph). Let's say the fire truck moves at 10 m/s (22.4 mph), and the hose is still spraying water. How fast will the water come out of the hose? For you (on the

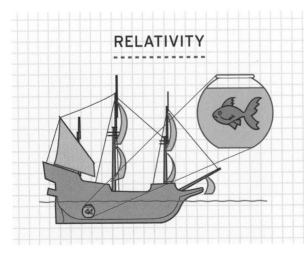

RELATIVITY

fire truck) it comes out at 10 m/s. For someone on the ground it comes out at 5 m/s + 10 m/s = 15 m/s (33.6 mph). To get this figure, the speed of the fire truck is added to the speed of the water coming out of the hose.

Einstein thought about riding the world's fastest fire truck. If he had a flashlight and a water hose on this fire truck, what would he see? Imagine he rides the fire truck at 150,000,000 m/s (336,000,000 mph). What would be the speed of the water coming out of the hose? It would be 150,000,005 m/s.

LIGHT IS SPECIAL

What about the light from Einstein's flashlight? Using the same logic as above, you might expect its speed to be about 450,000,000 m/s (1,000,000,000 mph), but this is wrong. The speed would be 299,792,458 m/s because the speed of light in empty space never changes. This would be the same for Einstein on the fire truck, and the same for someone on the ground.

How is this possible? It took a long time for Einstein to work out the answer. After a lot of careful thought, he realized that to make sure everyone agreed on the speed of light, the common understanding of relativity had to change. He discovered that as objects speed up, three things happen:

- Their clocks appear to run slower from the point of view of someone not moving.
- Their length appears to be shorter from the point of view of someone not moving.
- Their mass appears to increase from the point of view of someone not moving.

This was difficult for people to accept at first, but every single experiment that has ever been done has confirmed these three facts. Imagine you flew to the nearest star, Proxima Centauri, in a spaceship at half the speed of light. The journey would take 8.8 years from the point of view of someone on Earth, but you would only age 7.6 years on the spaceship.

ALBERT EINSTEIN

EXPERIMENT: GRAVITY AND GENERAL RELATIVITY

ADULT SUPERVISION REQUIRED

Earth tries to follow a straight line as it orbits the sun, but the sun turns this line into a circle. This experiment helps you understand how.

YOU WILL NEED:

- A makeshift trampoline (see the box below) or an actual trampoline, if you have one
- Marbles
- 2 heavy round stones
- Sheet of cardboard
- Sticky notes

WHAT TO DO:

1. Mark the edges of the trampoline like a clock, using sticky notes.

2. Fold the cardboard in half to make a chute for rolling marbles.

CURVED "SPACE"

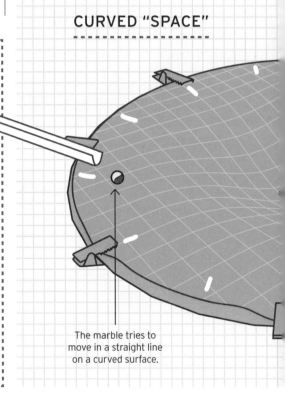

Cardboard chute

The marble tries to move in a straight line on a curved surface.

MAKE YOUR OWN SHEET OF FABRIC

- Plastic hoop
- Sheet of spandex or other stretchy fabric bigger than the hoop
- 8 strong binder clips
- 4 chairs

WHAT TO DO:

Get an adult to help. Gently stretch the fabric over the hoop. Attach the fabric to the hoop using clips, making sure the clips are spaced evenly around the hoop. Support the hoop on the four chairs, so that the center is at least 30 cm (about 1 ft.) above the floor.

3. Stand outside the trampoline at the 6 o'clock position and roll marbles toward the 9 o'clock, 12 o'clock, and 3 o'clock positions, using the cardboard chute. The marbles should follow a straight line over the surface of the trampoline.

4. Gently place one of the heavy stones in the center of the trampoline.

5. Repeat step 3 again. The marbles will no longer trace straight lines; instead, they will follow curved lines.

6. Roll the marbles at different angles and see if you can get them to the other side of the trampoline by going around the stone.

7. Place the two stones about 50 cm (1.5 ft.) away from each other on the trampoline. Can you make a marble trace a figure-eight shape around them?

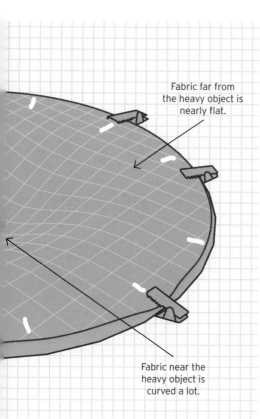

Fabric far from the heavy object is nearly flat.

Fabric near the heavy object is curved a lot.

WHAT HAPPENS?

Einstein's theory of relativity states that gravity is not a force; it is a bending of space (and time). When the marbles roll across the surface of an empty trampoline, they follow straight lines. Adding the heavy stone changes the shape of the surface of the trampoline and makes it curved. When the marbles move on the curved trampoline surface, they follow as much of a straight line as possible, but because the surface is curved, this line must also be curved.

Earth travels around the sun because the sun is so massive it curves space around it. If the sun weren't there, Earth would continue to travel through space in a straight line. The curving of space makes Earth move in a big circle, and this makes it look like there is a gravitational force pulling Earth toward the sun. Is gravity a force or curved space? It can be seen as either.

LEARN ABOUT: VERY FAST SPACESHIPS

The distances in space are very large. The fastest anything can travel is the speed of light, and even then, it would take 4.4 years to reach the star closest to Earth. To reach the next galaxy would take even longer— about 25,000 years, in fact.

THE EFFECTS OF RELATIVITY

Relativity has three main effects on very fast spaceships:

- Time slows down onboard a very fast spaceship.
- The length of a very fast spaceship shrinks.
- The mass of a very fast spaceship increases.

These are all changed by the same amount, known as the gamma factor. For example, if a spaceship was traveling so fast that its onboard clock was going at half the normal speed, then it would also appear to be half as long as normal and twice as massive as normal.

RELATIVITY AND SPACESHIPS

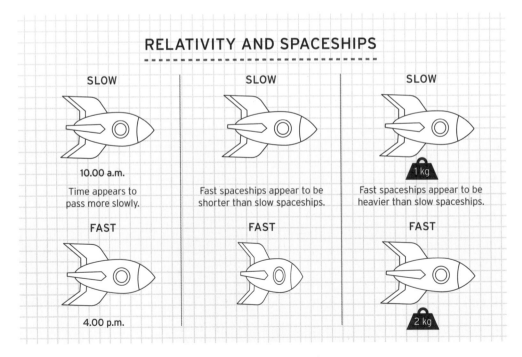

SLOW	SLOW	SLOW
10.00 a.m.		1 kg
Time appears to pass more slowly.	Fast spaceships appear to be shorter than slow spaceships.	Fast spaceships appear to be heavier than slow spaceships.
FAST	FAST	FAST
4.00 p.m.		2 kg

SPACE BREAK

Suppose you are taking a break on a space mission, stopping at a local planet. You see three spaceships flying past. You know that when they were made, they were 1,000 m long, they weighed 100 tonnes, and they had a beacon on the tail that flashed once every second. Can you arrange these spaceships in order of speed?

- The first spaceship appears to be 250 m long.
- The second spaceship appears to have a mass of 500 tonnes.
- The beacon on the third spaceship appears to flash once every 3 seconds.

COLOR AND SPEED

Another effect of relativity is that colors change depending on how fast a spaceship moves. This is because as the distances get smaller and times slow down for very fast spaceships, the wavelength, and therefore color, of the light coming from the spaceships also changes. Spaceships look bluer when they are moving toward you, and redder when they are moving away from you. The faster they move, the more blue or red they look. This is called redshift and blueshift. It's similar to how the pitch of an ambulance siren appears to rise as the ambulance approaches, and fall as the ambulance drives away.

SPOTTING SPACESHIPS

You wait around and see some more spaceships, which are white when they are at rest. To your left you see spaceship A, which is slightly blue, and spaceship B, which is very red. To your right you see spaceship C, which is extremely blue, and spaceship D, which is white. Which spaceship is moving the fastest with respect to you? For a pilot in spaceship C, which other spaceship would look the most colorful (most blue or most red), and which would look the least colorful?

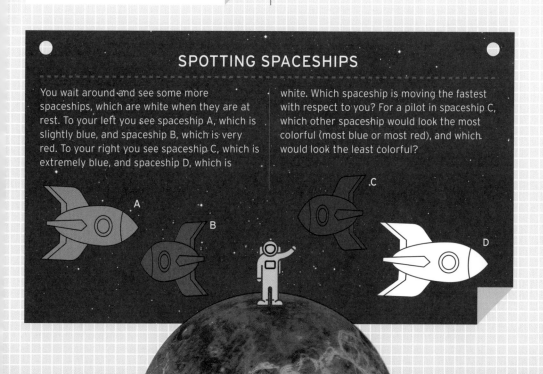

DISCOVER: EARTH AND OUR SOLAR SYSTEM

For thousands of years, people have looked up at the sky at night, fascinated by the stars and planets. What makes them move the way they do? Why do the stars always have the same arrangement in the sky? It took centuries to answer these questions and find Earth's place in the universe.

STARRY NIGHT

On a clear night, thousands of stars are visible in the sky. Wait an hour, and it will be clear that their positions in the sky have changed. All the stars seem to move in the same way, rotating as if they are on a giant sphere. There's one star that doesn't seem to move—Polaris, or the North Star.

Ancient Greek astronomers thought that the stars were on a giant spherical shell called the aether, with Polaris at the top. They thought that the aether rotated, while Earth stayed still, and that's why the stars seemed to move in the sky. In fact, Earth rotates around its axis, and this makes it look like the stars move. It took hundreds of years for Earth's rotation to be proven and accepted because it seems natural to think that Earth doesn't rotate—its movement is not discernable by the human senses alone.

Science is about making sense of the world by creating models. The models that scientists make should be as simple as possible. A complex model of the world with a lot of extra details, or assumptions, is likely to be wrong. If many details can be replaced with one detail, the model will be simpler and more likely to be correct. A model where all the stars move in the same way, but millions of miles away from each other, involves thousands of assumptions. A model where only Earth rotates is much simpler.

THE SUN AND THE MOON

The most obvious changes in the sky are the positions of the sun and the moon. Every day the sun and moon appear to move across the sky. This makes sense if the sun and moon move around Earth, or if Earth rotates on its axis. A model where Earth rotates is simpler than a model where both the sun and moon move separately.

Over the course of a year, the sun seems to appear higher in the sky in summer and lower in the sky in winter. This makes sense if the sun moves in a big circle around Earth, or Earth orbits the sun. Looking at a globe makes it clear that the axis of Earth is at an angle. It's this angle that makes the sun appear higher and lower in the sky, depending on the time of year.

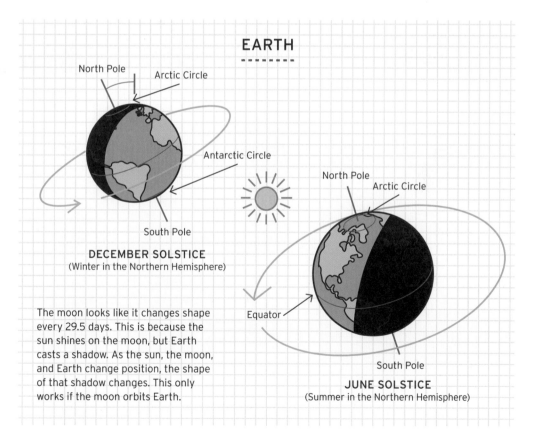

EARTH

North Pole
Arctic Circle
Antarctic Circle
South Pole

DECEMBER SOLSTICE
(Winter in the Northern Hemisphere)

North Pole
Arctic Circle
Equator
South Pole

JUNE SOLSTICE
(Summer in the Northern Hemisphere)

The moon looks like it changes shape every 29.5 days. This is because the sun shines on the moon, but Earth casts a shadow. As the sun, the moon, and Earth change position, the shape of that shadow changes. This only works if the moon orbits Earth.

EXPERIMENT: HOW THINGS LOOK FROM EARTH

The solar system is made up of the sun, eight planets, and thousands of other objects. They all move around, pulled by the force of gravity. It's easy to work out where each of the stars will be at a given time, but planets are different.

MAKING MAPS

About 500 years ago, scientists knew that Earth rotated on its axis, and that the moon orbited Earth, and they thought that the sun also orbited Earth. Even so, there were a few objects in the sky that seemed to move in strange ways against the background of the stars. For centuries, scientists kept track of these objects and called them planets, which means "wanderers" in Greek, because they would move forward for a while, then move backward, then move forward again.

DE MOTIB. STELLÆ MARTIS

PATH OF MARS

Scientists made a map of how the planet Mars moves from the point of view of Earth. They assumed that the sun orbited Earth, and that Mars orbited the sun. You can see what they drew on the left, with Earth at the center. You can trace the path of Mars with your finger from 1580 to 1596. You will find that it seems to move forward most of the time, but backward when it goes through a small loop.

YOU WILL NEED:
• **Spiral design sets**

WHAT TO DO:

If the pattern on the left looks familiar, that's because it is similar to the pattern produced by a spiral design set. The set works by making one circle (or other shape) move along the path of another, and it can be used to show how the orbit of Mars looks from Earth. Earth will be at the center of the large wheel, and the center of the small wheel represents the sun.

1. Begin to draw using the spiral design wheel. As you move the small wheel around, its center will always be the same distance from Earth.

2. Mars is about 1.5 times farther away from the sun than Earth is, so, for Mars, find a point that is 1.5 times farther from the center of the small wheel compared to the distance from Earth to the sun.

3. As you draw, you should find that you can make a pattern that looks like the one on page 142. Try different wheels to see how orbits of different planets would look from Earth. For example, Venus is about 0.75 times closer to the sun than Earth is.

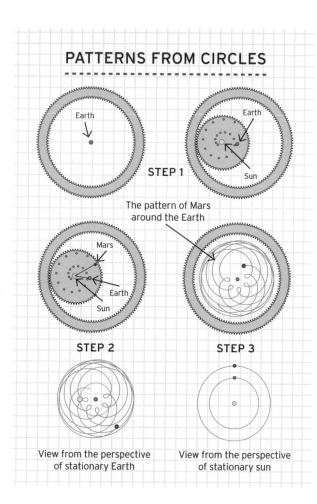

PATTERNS FROM CIRCLES

STEP 1

Earth

Earth

Sun

The pattern of Mars around the Earth

STEP 2

Mars

Earth

Sun

STEP 3

View from the perspective of stationary Earth

View from the perspective of stationary sun

WHAT HAPPENS?

The beautiful patterns reflect the way planets' orbits appear from the point of view of Earth. The view from the sun is much simpler because each planet looks like it travels in a circle. By investigating these patterns, scientists realized that the sun is at the center of our solar system, and Earth is just another planet. The orbits of the planets are not quite circles, so the patterns are a bit more complex, but they will keep tracing out these patterns for billions of years.

LEARN ABOUT: MEET THE PLANETS

Since ancient times, humans have known about five planets (other than Earth). These are visible at night without telescopes, and each one is slightly different—for example, Saturn is less dense than water and Jupiter has 79 moons.

THE PLANETS OF OUR SOLAR SYSTEM

To complete this puzzle, match the properties of the different planets by placing check marks and Xs. For example, out of the five planets, Saturn is farthest from the sun. This has been added to the big grid as a check mark, and Xs have been added to exclude the other planets. Use the remaining clues to add check marks and Xs until the grid is complete. You can then fill in the table below to find out about all these planets.

- The hottest planet has runaway global warming.
- The largest planet has a giant storm.
- The planet with a longer day than a year is closest to the sun.

- Mars is the red planet.
- Saturn has rings around its equator.
- The largest planet is Jupiter.
- The hottest planet is the second-closest to the sun, and the second-hottest planet is the closest to the sun.
- Mars is farther from the sun than Venus.
- The largest planet is the second-farthest from the sun, and the second-largest planet is the farthest from the sun.
- Mercury is the closest planet to the sun.
- Venus is larger than both Mercury and Mars.
- The smallest planet is Mercury.
- Saturn is colder than Jupiter, and Jupiter is colder than Mars.

	Distance from sun (km)	Surface temperature (C)	Radius (km)	Unique feature
Jupiter				
Mars				
Mercury				
Saturn				
Venus				

Of course, the invention of the telescope has prompted the discovery of more objects in our solar system. These include the planets Uranus and Neptune, the dwarf planet Pluto, the asteroid belt, and comets. So far, eight planets, five dwarf planets, 181 moons, 3,083 comets, and more than half a million asteroids have been found.

		Jupiter	Mars	Mercury	Saturn	Venus	58,000,000 km	108,000,000 km	228,000,000 km	779,000,000 km	1,434,000,000 km	-150°C	-145°C	-63°C	427°C	462°C	2,440 km	3,390 km	6,052 km	58,232 km	69,911 km	Giant storm	Red surface	Rings around equator	Runaway global warming	Day is longer than year
Name	Jupiter	✓	✗	✗	✗	✗					✗															
	Mars	✗	✓	✗	✗	✗					✗															
	Mercury	✗	✗	✓	✗	✗					✗															
	Saturn	✗	✗	✗	✓	✗	✗	✗	✗	✗	✓															
	Venus	✗	✗	✗	✗	✓					✗															
Distance from the sun	58,000,000 km				✗		✓	✗	✗	✗	✗															
	108,000,000 km				✗		✗	✓	✗	✗	✗															
	228,000,000 km				✗		✗	✗	✓	✗	✗															
	779,000,000 km				✗		✗	✗	✗	✓	✗															
	1,434,000,000 km	✗	✗	✗	✓	✗	✗	✗	✗	✗	✓															
Surface temperature	-150°C											✓	✗	✗	✗	✗										
	-145°C											✗	✓	✗	✗	✗										
	-63°C											✗	✗	✓	✗	✗										
	427°C											✗	✗	✗	✓	✗										
	462°C											✗	✗	✗	✗	✓										
Radius	2,440 km																✓	✗	✗	✗	✗					
	3,390 km																✗	✓	✗	✗	✗					
	6,052 km																✗	✗	✓	✗	✗					
	58,232 km																✗	✗	✗	✓	✗					
	69,911 km																✗	✗	✗	✗	✓					
Unique feature	Giant storm																					✓	✗	✗	✗	✗
	Red surface																					✗	✓	✗	✗	✗
	Rings around equator																					✗	✗	✓	✗	✗
	Runaway global warming																					✗	✗	✗	✓	✗
	Day is longer than year																					✗	✗	✗	✗	✓

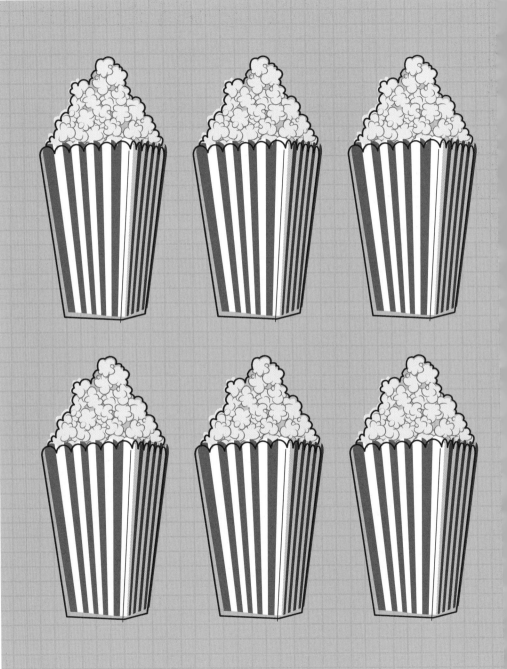

ANSWERS

ANSWERS

P. 12 THE FIRST POP

When you add popped popcorn to water, the water level hardly rises. This is because the popcorn absorbs the water. It can absorb the water because it has large pockets of air inside that can be filled with water. The density decreases because the popped popcorn is mostly pockets of air.

P. 20 DENSITY

These answers are approximations. Your answers will probably vary from these slightly.

The mass of an unpopped kernel is about 0.16 g, and the mass of a popped piece of popcorn is about 0.14 g, which gives a difference of around 12%.

Eighty-three unpopped kernels make the water level go up by 10 ml. The mass of 83 unpopped kernels is 83 x 0.16 g = 13.3 g. This gives a density of 13.3 g ÷ 10 ml = 1.33 g/ml.

Twenty-five pieces of popped popcorn occupy 125 ml of volume. The mass of 25 pieces of popped popcorn is 25 x 0.14 g = 3.5 g. This gives a density of 3.5 g ÷ 125 ml = 0.028 g/ml.

The ratio of these densities is 1.33 ÷ 0.028 = 47.5.

Because popcorn absorbs some of the water, the water occupies some of the space inside the popcorn. If the same amount of mass is occupying a much larger volume, then it must be more spread out, and there must be large gaps in the structure.

When selling popcorn, it is best to sell it unpopped. When it is popped, it has roughly the same mass, but it takes up much more space, which makes it a lot more expensive to move around and store.

P. 21 HEAT TRANSFER

1. *It's time for dinner, and your friend wants to make pasta. They put some pasta in a pot of cold water and cook it on the stove.* Conduction from the stove to the pot. Convection as heat moves through the water.

2. *It's a hot day, and you're drinking lemonade. A piece of ice falls out of your glass and onto the ground. The ice is melted by the sun.* Radiation from the sun to the ice and convection from the ground to the ice.

3. *You're eating chocolate, but you are taking your time. It melts in your hands.*
Conduction from your hand to the chocolate.

4. *You start to run a hot bath, and you can see water vapor rising from the surface of the water.* Convection within the water. Convection of water vapor rising.

5. *You're camping, and you warm yourself by a campfire.*
Radiation from the fire to you. Convection as the air around the fire heats up.

6. *You sprain your ankle, and the school nurse brings you an ice pack, which you apply to the sprain.*
Conduction from your ankle to the ice pack.

7. *Your parents drive you to school on a cold day. They turn on the heater in the car to blow warm air into the car.*
Convection as the warm air mixes with the cold air.

P. 24 TURNING WATER TO STEAM

These answers are approximations. Your answers will probably vary from these slightly.
• Mass of kettle before boiling: 1,636 g
• Mass of kettle after boiling: 1,631 g
• Mass of water lost: 5 g
• Volume of water lost: 5 cm³
• Diameter of the spout: 2 cm
Area of the spout:
$\pi \times 2 \text{ cm} \times 2 \text{ cm} \div 4 = 3.14 \text{ cm}^2$

- It took 0.5 seconds for the steam to rise 50 cm, so the speed of steam was:
  50 cm ÷ 0.5 s = 100 cms^{-1}
- The volume of steam produced per second was therefore:
  3.14 cm^2 × 100 ms^{-1} = 314 cm^{3}s^{-1}
- The amount of time between steam first being produced and the water boiling was 10.5 seconds.
- The volume of steam produced was:
  314 cm^{3}s^{-1} × 10.5 s = 3,297 cm^3
- Therefore the volume of steam was about 660 times larger than the volume of water boiled to make the steam.

P. 28 THE PRESSURE OF THE POP

Suppose the amount of steam that escapes a kernel is 0.02 g.

Suppose that the volume of the kernel is 0.4 cm^3.

The temperature is 176ºC.

The pressure is given by:

Pressure = Amount of steam per kernel (g) × Temperature (C) ÷ Volume of kernel (cm^3)

Putting these values in gives:

Pressure = 0.02 g × 176ºC ÷ 0.4 cm^3 = 8.8

So the pressure inside the kernel is about 9 times more than atmospheric pressure.

P. 30 BALANCING EQUATIONS

- A popcorn kernel is heated. The amount of steam (which is a gas) inside it increases, and so does the temperature. The volume of steam stays the same.

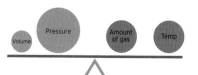

Volume x Pressure = Amount of gas x Temperature
Temperature and amount of gas increase, pressure increases, seesaw is balanced.

- The air inside a hot-air balloon cools down. The pressure of the air does not change, and the amount of air stays the same.

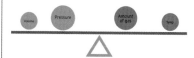

Volume x Pressure = Amount of gas x Temperature
Temperature decreases, volume decreases, seesaw is balanced.

- You blow up a balloon and add more air to it. The temperature of the air stays the same.

Volume x Pressure = Amount of gas x Temperature
Amount of gas increases, pressure and volume increase, seesaw is balanced.

- You open a bottle of soda and notice a lot of air escapes from the bottle. The temperature of the air stays the same.

Volume x Pressure = Amount of gas x Temperature
Amount of gas decreases, pressure decreases, seesaw is balanced.

P. 34 PROPERTIES OF SOLIDS, LIQUIDS, AND GASES

1. a) The molecules have no overall pattern, and don't tend to stay together. (g)

1. b) The molecules are well organized in a repeating pattern. (s)

1. c) The molecules have no overall pattern, but they are generally close together. (l)

2. a) The molecules touch each other. (s)

2. b) There is a lot of space between the molecules. (g)

2. c) The molecules are close together, but can slide over each other. (l)

3. a) The molecules can only vibrate in place. They cannot move around. (s)

3. b) The molecules are free to move around each other, but tend to stay close to other molecules. (l)

3. c) The molecules are free to move anywhere, and move quickly in all directions. (g)

4. a) Examples include water, oil, and lava. (l)

4. b) Examples include air, steam, and helium. (g)

4. c) Examples include sand, wood, and rubber. (s)

5. a) The overall shape matches the shape of its container, but its volume stays the same. (l)

5. b) The overall shape stays the same unless it gets deformed. (s)

5. c) The overall shape matches the shape of its container, but its volume will expand to fill its container. (g)

6. a) When heated, it eventually boils. (l)

6. b) When heated, it expands or its pressure increases. (g)

6. c) When heated, it expands slightly, and eventually melts. (s)

7. a) If you hit it with a hammer, nothing significant happens. (g)

7. b) If you hit it with a hammer, only the substance that touches the hammer will move. You might make waves appear. (l)

8. c) If you hit it with a hammer, the whole substance will move. (s)

P. 45 ELECTRICITY AND MAGNETISM

1.

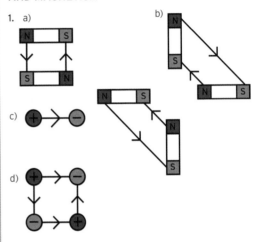

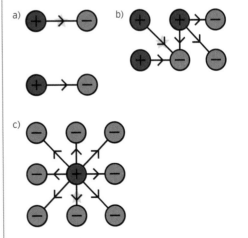

2. The arrows show the direction the charged objects would move in, if placed where the stars are.

3. The arrows show the direction the compass needles would face, if placed where the stars are.

a)

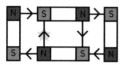

b)

c)

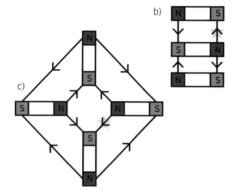

P. 56 HOW FAST ARE WAVES?

1. Speed of water waves in the ocean (about 2 m/s [4.5 mph])

2. Speed of sound waves in the air (343 m/s [767 mph])

3. Speed of sound waves in water (1,482 m/s [3,315 mph])

4. Speed of p waves in earthquakes (13,992 m/s [31,300 mph])

5. Speed of light waves (299,792,458 m/s [670,616,629 mph])

P. 57 BEYOND VISIBLE LIGHT

a) Scanning airport luggage for dangerous materials (X-rays)
b) Studying the structure of metals at a very small scale (X-rays)
c) Making medical images of the inside of human bodies (X-rays)
d) Killing microbes on medical equipment (Ultraviolet)
e) Detecting forgeries of banknotes (Ultraviolet)
f) Sensing when a human is within a few feet (Infrared, visible light, gamma rays)
g) Making thermal images of heat (Infrared)
h) Night-vision cameras (Infrared)
i) Heating food in a microwave oven (Microwaves)
j) Cell phone and Wi-Fi signals (Microwaves)
k) Transmitting messages over very long distances (Radio waves)
l) Global Positioning System (GPS) signals (Radio waves)

P. 64 CIRCUITS

With a power rating of 800 W and a potential difference of 110 V, the current is given by:

Current = Power ÷ Potential difference
Current = 7.3 A

The resistance is given by:

Resistance = Potential difference ÷ Current
Resistance = 15 ohm

To decrease the power by a factor of 3, you need to decrease the current by a factor of 3. This means for the same potential difference, you need to increase the resistance by a factor of 3. So the resistance should be 45 ohm.

P. 65 DIFFERENT POWER PLANTS

1. Coal-burning power plant (Nonrenewable)

2. Solar power plant using solar cells (Renewable)

3. Wind power plant (Renewable)

4. Oil-burning power plant (Nonrenewable)

5. Hydroelectric (falling water) power plant (Renewable)

6. Gas-burning power plant (Nonrenewable)

7. Solid waste-burning power plant (Renewable)

8. Tidal power plant (using tides in the ocean) (Renewable)

9. Nuclear power plant (Nonrenewable)

10. Geothermal power plant (using Earth's heat) (Renewable)

P. 74 SENDING SECRET CODES WITH PHOTONS

1. Bob sends OKAY LETS GO ON TUESDAY

2. Bob sends HI ALICE THIS IS BOB

3. Alice sends HELLO

P. 82 SPEED GRAPHS

1. The acceleration of the car is (22 m/s - 13 m/s) ÷ 10 seconds = 9 m/s ÷ 10 seconds = 0.9 m/s each second. The average speed of the car is (22 m/s + 13 m/s) ÷ 2 = 17.5 m/s.

2. The acceleration of the bus is:
10 m/s ÷ 6 seconds = 1.67 m/s each second.
The distance the bus travels is given by Speed × Time. The average speed is 5 m/s, so the distance = 5 m/s × 6 seconds = 30 m.

3. The acceleration of the truck is:
6 m/s ÷ 5 seconds = 1.2 m/s each second.
The distance the car travels is the area under the blue lines. During each period of acceleration, this is: 10 × 5 ÷ 2 = 25 m. During each period of constant speed, the distance is: 10 × 5 = 50. There are four periods of acceleration and two periods of constant speed. Adding that all up gives:
(4 × 25) + (2 × 50) = 200 m.

Using the same method for the truck: During each period of acceleration 6 × 5 ÷ 2 = 15 m. During the period of constant speed, the distance is: 6 × 35 = 210 m. There are two periods of acceleration and one period of constant speed, so the distance is: (2 × 15) + (1 × 210) = 240 m.

The truck travels farther than the car.

P. 88 GRAVITY AND THOUGHT EXPERIMENTS

Aristotle thinks that heavier objects accelerate more under gravity, so he thinks that the heavier object will hit the ground before the lighter object. Galileo thinks that all objects accelerate the same amount under gravity, so he thinks both objects will hit the ground at the same time.

Galileo suggests making the piece of string connecting the two objects very small. If the piece of string is short enough, the two objects together are the same as one very heavy object. According to Aristotle, this should fall even faster than the heavy object. The light object cannot have the effect of making the heavy object fall both slower and faster at the same time. It's absurd that the length of the string would determine how the light object affects the heavy object. The only way out of the paradox is that all objects must fall with the same acceleration.

P. 94 PLAYING POOL

These are possible solutions, although there may be more.

1.

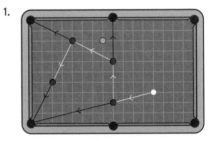

2.

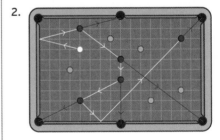

4. In these answers, north and east are positive, and south and west are negative.

- **East-West:** 6 units (cue ball before) + 0 units (red ball before) = 5 units (cue ball after) + 1 unit (red ball after)
  North-South: 0 units (cue ball before) + 0 units (red ball before) = -2 units (cue ball after) + 2 units (red ball after)

- **East-West:** 0 units (cue ball before) + 0 units (red ball before) = 1 unit (cue ball after) - 1 unit (red ball after)

North–South: -3 units (cue ball before) + 0 units (red ball before) = -1 unit (cue ball after) - 2 units (red ball after)

- **East–West:** -2 units (cue ball before) + 0 units (red ball before) = -1 unit (cue ball after) - 1 unit (red ball after)
North–South: -3 units (cue ball before) + 0 units (red ball before) = 0 units (cue ball after) - 3 units (red ball after)

P. 100 ANGULAR MOMENTUM
When the ice-skater brings his arms in, he is less spread out. That means if nothing else changed, his angular momentum would decrease. Angular momentum is conserved, so something else must change to keep it the same value. His mass cannot change, so the speed at which he spins must increase. If he brings his arms in, he will spin faster. Using the same logic, if he moves his arms out, he will spin more slowly.

When the ice-skater picks up his daughter, he increases his overall mass. His angular momentum stays the same, but increasing mass has the effect of increasing angular momentum, so something else must change to keep his angular momentum constant. That means he spins more slowly after picking up more mass.

Changing Earth's rotation
The villain can either make Earth pancake shaped to spread out the existing mass from the axis, or add lead to the surface of Earth to make Earth more massive. Both of these would cause Earth to slow down its spin to conserve angular momentum. Making Earth broomstick shaped or hollow would cause it to spin more quickly.

There are 365 × 24 = 8,760 hours in a year. Working for 8 hours per day, for 5 of every 7 days means people would work 365 × 8 × 5 ÷ 7 = 2,086 hours per year. If a day becomes 30 hours long, there would be 8,760 ÷ 30 = 292 days in a year. Working for 10 hours per day, for 5 of every 7 days means people would work 292 × 10 × 5 ÷ 7 = 2,086 hours per year.

The villain could make the day longer, but he cannot change the number of hours in a year. Either way, most people will still work just over 2,000 hours per year.

P. 107 FRICTION AND TRUCKS
Escape lane
Average speed = 15 m/s
Distance = 50 m
Time available = 50 m ÷ 15 m/s = 3.3 s
Maximum deceleration = (3 + 1) m/s each second = 4 m/s each second
Maximum change in speed = 4 × 3.3 = 13.3 m/s
This is not enough of a change in speed to stop the truck.

Gravel surface
Average speed = 15 m/s
Distance = 175 m
Time available = 175 m ÷ 15 m/s = 11.7 s
Maximum deceleration = (2 + 1) m/s each second = 3 m/s each second
Maximum change in speed = 3 × 11.7 = 35 m/s
This is enough of a change in speed to stop the truck.

If the driver chose the gravel surface, with deceleration of friction of 2 m/s each second, then the total deceleration would be 3 m/s each second, and the time taken would be 10 seconds. This means the final distance would be 150 m, and the driver would have enough gravel to stop the truck completely.

Therefore, she should choose the gravel surface.

P. 112 NUCLEAR POPCORN

Here is a typical graph used to find the half-life of popcorn:

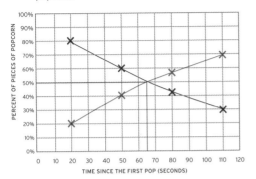

- Fraction of unpopped kernels
- Fraction of popped kernels
- Half-life = 66 seconds

P. 115 INSIDE AN ATOM

The volume of the proton is $10^{-15} \times 6 \times 10^{-31}$ m³ = $6 \times 10^{-15-31}$ m³ = 6×10^{-46} m³

Density = Mass ÷ Volume = $(1.7 \times 10^{-27}$ kg) ÷ $(6 \times 10^{-46}$ m³) = 0.28×10^{19} kg m⁻³

For the electron: Density = Mass ÷ Volume = 9.1×10^{-31} kg ÷ 6×10^{-31} = 1.5 kg m⁻³

That means the nucleus is about $0.28 \times 10^{19} \div 1.5$ = 1,900,000,000,000,000,000 times more dense than the electron cloud. That's about 2 billion billion times more dense.

P. 121 RADIATION AND ORANGES

1. Each radiation source gives off a two orange equivalent dose per hour, so after one hour, they give a six orange equivalent dose.

After one day they give a 144 orange equivalent dose.

Assuming 365 days in a year, this gives a 52,560 orange equivalent dose per year, which is more than twice the recommended limit.

2. Adding a sheet of paper will stop the alpha radiation, so the dose would be $4 \times 24 \times 365$ = 35,040 orange equivalent dose per year. This is still above the recommended limit.

Adding a sheet of metal will stop the alpha radiation and beta radiation, so the dose would be $2 \times 24 \times 365$ = 17,520 orange equivalent dose per year. This is below the recommended limit.

Half of the gamma radiation can get past 1 cm (about 0.5 in.) of lead. That means a quarter can get through 2 cm (about 0.8 in.) of lead, and an eighth can get through 3 cm (about 1.2 in.) of lead. If you add 3 cm of lead, only a 2,190 orange equivalent dose per year will get through.

P. 127 ESCAPE VELOCITIES

Romulus has more mass then Earth, but is the same size as Earth, so it has a larger escape velocity than Earth.

Remus has the same mass as Earth, but it is larger than Earth, so it has a smaller escape velocity than Earth.

Vulcan has more mass then Earth, and it is smaller than Earth, so it has a much larger escape velocity than Earth.

P. 132 SPACE MISSIONS

1. Sputnik: (d)

2. Pioneer 1: (j)

3. Vostok 1: (h)

4. Mariner 2: (i)

5. Apollo 11: (b)

6. Voyager 1: (a)

7. Giotto: (c)

8. Mars Pathfinder: (g)

9. International Space Station: (e)

10. Hayabusa 2: (k)

11. Europa Clipper: (f)

P. 139 VERY FAST SPACESHIPS

Space Break

- The first spaceship has a gamma factor of 1,000 m ÷ 250 m = 4.
- The second spaceship has a gamma factor of 500 tonnes ÷ 100 tonnes = 5.
- The third spaceship has a gamma factor of 3 seconds ÷ 1 second = 3.
Therefore, the second spaceship is traveling the fastest, and experiencing the most extreme effects of relativity.

Spotting spaceships

- Spaceship C is extremely blue, so it is moving most quickly (and is moving toward you).
- From spaceship C's viewpoint, all other spaceships are moving toward it at different speeds. Since spaceship A is moving toward you from the other direction, it has the largest speed from the viewpoint of spaceship C, so it is the most colorful. It looks even more blue to spaceship C than spaceship C does to you. Spaceship B looks the least colorful to spaceship C because from your point of view, it is the only spaceship moving in the same direction as spaceship C.

P. 144 MEET THE PLANETS

	Distance from sun	Surface temperature	Radius	Unique feature
Jupiter	779,000,000 km	-145°C	69,911 km	Giant storm
Mars	228,000,000 km	-63°C	3,390 km	Red surface
Mercury	58,000,000 km	427°C	2,440 km	Day is longer than year
Saturn	1,434,000,000 km	-150°C	58,232 km	Rings around equator
Venus	108,000,000 km	462°C	6,052 km	Runaway global warming

INDEX

PICTURE CREDITS